Jet Grouting

Technology, Design and Control

Jet Grouting

Technology, Design and Control

Paolo Croce
Alessandro Flora
Giuseppe Modoni

CRC Press
Taylor & Francis Group
Boca Raton London New York

CRC Press is an imprint of the
Taylor & Francis Group, an **informa** business

A SPON PRESS BOOK

CRC Press
Taylor & Francis Group
6000 Broken Sound Parkway NW, Suite 300
Boca Raton, FL 33487-2742

First issued in paperback 2017

© 2014 by Taylor & Francis Group, LLC
CRC Press is an imprint of Taylor & Francis Group, an Informa business

No claim to original U.S. Government works
Version Date: 20140108

ISBN 13: 978-1-138-07627-3 (pbk)
ISBN 13: 978-0-415-52640-1 (hbk)

Visit the Taylor & Francis Web site at
http://www.taylorandfrancis.com

and the CRC Press Web site at
http://www.crcpress.com

To the memory of our fathers.

Contents

6 Design principles 123

7 Design examples 155

Preface

The jet grouting technique has become very popular all over the world as a practical means for solving several geotechnical problems. Sometimes, however, the inappropriate use of jet grouting has led to unsuccessful results raising relevant concern on its efficiency. Nevertheless, these failures usually stem from ignorance about the actual possibilities and limitations of the method rather than from the value of the method itself.

In fact, in order to manage a jet grouting project, it is essential to have detailed knowledge of the various technological procedures that could be used and, above all, to understand the possible effects of jet grouting on different natural soils. In recent years, this latter aspect has been subjected to thorough studies, which have substantially increased the knowledge of the complex jet–soil interaction phenomena. The results of these studies have recently provided enhanced and reliable methods to estimate the effects of jet grouting, relating them to both soil properties and treatment procedures. However, such studies are still mostly confined to the scientific literature, known only to a small academic community, and are not yet sufficiently widespread among practicing engineers.

Another crucial issue is the analysis of the so-called jet-grouted structures, that is, the various possible assemblages of jet-grouted columns conceived to provide a required geotechnical function. This topic is of paramount importance for providing rational design methods, which are urgently needed for establishing jet grouting as a fully reliable and mature technology. Again, rational solutions are nowadays available, yet not known to the majority of practicing engineers who still tend to rely on purely empirical *rules of thumb*, mostly proceeding on the basis of personal experience or employing some sort of trial-and-error procedure.

In order to try and fill these gaps, this book presents an overview of engineering practice and research activity on jet grouting, with the perhaps ambitious aim of putting them into a unique, interconnected and consistent framework. For such a purpose, the authors have tried to tie together the technological issues, the interpretation of the mechanisms taking place during jet grouting, the quantitative prediction of their effects, the design of jet-grouted structures and finally the procedures for controlling the actual jet grouting results.

The authors are well aware that they may fall short of achieving their aims for this book. Clearly, some of the topics dealt with could be further refined, while others may be added. Moreover, some of the proposed solutions may raise questions and require more debate with particular regard to the design methods. Finally in certain cases the authors have pointed out some problems but have not been able to provide the relevant answers yet. Therefore, the book may be seen as a work in progress dealing with an ever-changing technique for which criticisms, discussion and contributions are most welcome and will help to fill possible gaps.

As far as contributions are concerned, it is finally important to recall that information and data on applications are essential, mistakes being more important than successful applications. Therefore, close and collaborative interactions among researchers, designers, contractors and equipment manufacturers are required to enhance both construction procedures and design methods with the final goal of improving rational and justified confidence in jet grouting.

The book is the outcome of a long-lasting cooperation among the authors and is the result of interaction with many colleagues and friends. The authors are deeply indebted to all of them for sharing experience and knowledge.

Authors

Paolo Croce graduated in civil engineering at the University of Napoli, Federico II, in 1979. He earned his master's degree in civil engineering from the University of Colorado in 1982.

He is a professor of geotechnical engineering at the University of Cassino and Southern Lazio, Italy, where he is currently teaching courses on soil mechanics and on slope stability. He is a member of the American Society of Civil Engineers (ASCE) and of the Italian Geotechnical Society (Associazione Geotecnica Italiana [AGI]), where he is a member of the AGI Committee for the Guidelines on Jet Grouting. He is also a convenor of the European Work Group on Ground Improvement (EG14 [CEN/TC 250]) and a member of the European Committee on Execution of Special Geotechnical Works (CEN/TC 288).

Professor Croce has 30 years of professional experience regarding earth and rockfill dams, tunnelling, foundations, earthworks and landslide control. He has authored several articles in geotechnical journals and conferences.

Alessandro Flora graduated in civil engineering at the University of Napoli, Federico II, in 1989. He earned his PhD in geotechnical engineering jointly from the University of Napoli, Federico II, and Roma La Sapienza in 1995. He is a professor of geotechnical engineering at the University of Napoli, Federico II, where he is currently teaching ground improvement and geotechnical works. Dr Flora is the author of approximately 100 research articles published in major international journals, in conference proceedings and in books. He is a member of the Italian Geotechnical Society (Associazione Geotecnica Italiana [AGI]), where he is a member of the AGI Committee for the Guidelines on Jet Grouting. He is also a member of the following International Society of Soil Mechanics and Geotechnical Engineering (ISSMGE) committees: TC211: Ground Improvement (Italian delegate) and TC301: Preservation of Monuments and Historic Sites (secretary). He is also the secretary of the European Work Group on Ground Improvement (EG14 [CEN/TC/250], Eurocode 7).

Dr Flora has been involved as a geotechnical consultant in the design and construction of earth dams, the foundations of large buildings, large underground excavations, tunnels and the stabilisation of landslides.

Giuseppe Modoni graduated in civil engineering at the University of Napoli, Federico II, in 1994. He finished his PhD in geotechnical engineering jointly from the University of Rome and the University of Napoli in 1999. He is a professor of geotechnical engineering at the University of Cassino and Southern Lazio (Italy), where he is currently teaching soil mechanics, foundation engineering and ground improvement. He is the author of research articles published in main international journals, in conference proceedings and in books. He is a member of the Italian Geotechnical Society (Associazione Geotecnica Italiana [AGI]), and is also a member of the AGI Committee for the Guidelines on Jet Grouting.

Dr Modoni has acted as geotechnical consultant in the design and construction of foundations of bridges and buildings, earth-retaining structures, tunnels and the stabilisation of landslides.

List of symbols

C coefficient expressing the efficiency of injection
c cohesion
$CV(x)$ variation coefficient of a statistical variable
C_μ parameter of the κ-ε turbulence model
d diameter of the nozzle
D_a average column diameter
D_d design value of column diameter
D_k characteristic value of column diameter
D_m mean diameter
D_{shaft} diameter of shafts
D_{tunn} diameter of tunnels
E Young modulus
E_n energy at the nozzle
E_p energy at the pump
E'_n specific energy (per unit length of column) at the nozzle
E'_p specific energy (per unit length of column) at the pump
F filling ratio
f_{bd} bonding strength between steel (or fibreglass) and jet-grouted material
f_{Jg} tensile strength of jet-grouted material
h hydraulic head
H_{lim} limit transversal force on jet grouting columns
j hydraulic gradient
K coefficient relating the Young modulus and the uniaxial compressive strength of jet-grouted material
k permeability coefficient
M number of nozzles
M_{ult} ultimate flexural moments of reinforced columns
n_g monitor rotation per each lifting step
N_{spt} number of blows in standard penetration tests
N_{ult} ultimate normal force of columns
N_γ, N_c, N_q bearing capacity factors for shallow foundations
p_0 pressure at the nozzle

p_{axis} dynamic pressure at the jet axis

$P_{f,s}$ accepted probability of failure for the single unit comprising the structures

$P_{f,u}$ accepted probability of failure for the single unit comprising the structures

P_f accepted probability of failure

p_g, p_w, p_a pressure of, respectively, grout (g), water (w) and air (a)

p_{out} pressure of the fluid outside the nozzle

P_{ULS} ultimate limit load at the tip of the jet column

q_c unit tip resistance of cone penetration tests

Q_g, Q_w, Q_a flow rates of grout (g), water (w) and air (a)

q_{lim} limit load of shallow foundation

q_u unconfined compressive strength of the jet-grouted material

$q_{u,d}$ design value of the uniaxial compressive strength

$q_{u,k}$ characteristic value of the uniaxial compressive strength

Q_{ULS} ultimate limit axial load of the jet-grouted column

r radial distance from the jet axis

$Re = \rho \cdot v_0 \cdot d_0 / \mu$ Reynold number at the nozzle

s spacing between columns

s_o spacing between columns at ground level

s_u undrained soil strength

$s(z)$ span between columns at different depths

$SD(x)$ standard deviation of statistical variables

SF_{up} safety factor against uplifting of bottom plugs

SF_{up}^* safety factor against uplifting of bottom plugs accounting for defects

S_{ULS} ultimate limit load at the shaft of the jet column

t thickness of panels

T_{ult} ultimate normal force of reinforcement

v_0 velocity of the injected fluid at the exit from the nozzle

v_{axis} mean component of velocity oriented along the jet axis

v_c penetration velocity of jets

V_g volume of grout injected for the unit length of the column

v_L limit penetration velocity of soil

V_m volume of injected grout per unit length of treatment

v_p, v_s propagation velocity of compression and shear waves

v_r average lifting speed of the monitor

W_c mass of injected grout per unit length of treatment

$W(x)$ hydrodynamic power of the jet at a distance x from the nozzle

x axial distance from the nozzle

x_{mean} mean value of a statistical variable

α azimuth angle of column axis

α_{bd} factor relating f_{bd} with q_u

α_E parameter accounting the influence of air shrouding on Λ

β deviation angle of columns from the prescribed directions

β_E coefficient relating the stiffness and the uniaxial compressive strength of jet-grouted material

Δs lifting step

Δt time interval per lifting step

ε dissipation rate (m^2/s^3)

γ_D partial safety factor for the diameter of columns

γ_M partial safety factor for the uniaxial compressive strength of jet-grouted material

$\gamma_s, \gamma_w, \gamma_{jg}$ unit weights of, respectively, water, soil and jet-grouted material

κ coefficient defining the turbulent kinetic energy

Λ parameter defining the profile of velocity outside the nozzle

$\Lambda^*, \Lambda_w^*, \Lambda_g^*$ parameter Λ for single fluid, and values computed for water and grout

λ_E energetic efficiency of jet grouting

λ_V volumetric efficiency of jet grouting

μ viscosity of the generic injected fluid

μ_w, μ_g viscosity of, respectively, water and grout

$\mu_t = \rho C_\mu \dfrac{\kappa^2}{\varepsilon}$ turbulent kinematic viscosity of the injected fluid

ρ density of the generic injected fluid

ρ_w, ρ_g density of, respectively, water and grout

τ_L limit vertical stress on the shaft of jet columns

φ friction angle

ω rotational velocity of the monitor

Ω cement–water ratio by weight

Chapter 1

Introduction

1.1 GROUND IMPROVEMENT AND GEOTECHNICAL ENGINEERING

Ground improvement methods are becoming increasingly popular in geotechnical engineering to solve construction problems and to offer new design solutions. In fact, in recent years, significant industrial research has been carried out to develop and refine several ground improvement techniques. As a consequence, new technologies, materials and applications are regularly proposed on the market and offered to the practitioner as a convenient solution.

Ground improvement techniques are nowadays important arrows in the quiver of design solutions of geotechnical engineers, and there are few major projects carried out without using at least one of them. In fact, the availability of such new technologies has significantly widened the range of design alternatives. However, continuous technical evolution and rapidly expanding areas of application pose new problems to the geotechnical engineers, who are not always provided with relevant experience on this topic. In many cases the ground improvement issues are only roughly considered at the design stage, and their solution is largely left to the specialised contractor. As a result, the designer may partly lose the control of the whole geotechnical process.

It follows that ground improvement technologies are often used without the same design care that is usually adopted for other traditional technologies – simply adopting them because they may help or even because 'more is better' – thereby a bit like adding suspenders to a belt. However, by following such an approach, the rational base of the scientific method, which should always be the guiding light at the base of a correct design, may be substantially betrayed.

Whatever the case, ground improvement techniques often form a grey part of design, where unforeseen problems may hide. This is typical of all situations in which technology has a turbulent growth, and basic research, education and knowledge dissemination have not been able to keep pace. Fortunately, in the last decade or so, there have been an increasing number of scientific articles and some books devoted to ground improvement. In addition, university courses specifically focused on ground improvement are spreading worldwide, and even the most skeptical geotechnical engineers have recognised that these technologies, if correctly designed and controlled, are a providential opportunity and not just a new problem. Indeed, it is all a matter of knowledge and confidence, and the problem is not the technology but its use or misuse—a perennial question common not only to geotechnical engineering.

1.2 BRIEF HISTORY OF JET GROUTING

Among all ground improvement techniques, jet grouting has certainly acquired a special place. In fact, nowadays, everybody knows that jet grouting can provide convenient solutions to most geotechnical problems. However, this powerful means is the result of a long-standing sequence of technical developments, which is still ongoing, and the contribution of jet grouting pioneers should be properly recognised.

Greenwood (Croce and Flora 2001) reports on a pioneering experience performed in 1962 by Cementation Co., Ltd., to create a cut-off wall in Pakistan, as briefly reported in the discussion of a symposium on Grouts and Drilling Muds in Engineering Practice (British National Society of ISSMFE 1963). However, it is commonly recognised that the jet grouting technique was originated in Japan (Nakanishi 1974).

In fact, in the late 1960s, experience on the use of high-speed jets for cutting rock and rock-like materials (Farmer and Attewell 1965) inspired a group of Japanese specialists to investigate their use as a ground improvement tool. They envisaged the possibility of injecting fluid binders within previously drilled boreholes to erode and mix in place the soil, with the aim to produce bodies of cemented material within the soil. In the subsequent decades, the technique has developed significantly and is now being widely used around the world. At present, jet grouting is arguably the most common ground improvement method.

In the first patented version, known as the Chemical Churning Pile (CCP) (Miki 1973; Nakanishi 1974), chemical binders were used, but these products were soon replaced by water-cement grouts. An evolution of the technique, known as the Jumbo Special Pile (JSP), was developed after some years by the same group, aimed at improving the dimensions of the jet-grouted elements (Xanthakos et al. 1994). It consisted of shrouding the jet of the grout with a coaxial stream of compressed air to reduce the loss of energy in the former and to preserve its erosive power over a greater distance.

Almost contemporarily, a different system referred to as the 'Jet Grout' method was conceived by another group of specialists (Yahiro and Yoshida 1973; Yahiro et al. 1974), in which treatment consisted of eroding the soil with high-speed jets of water followed by filling the remoulded material with cement grout injected from a lower nozzle. Initially, the nozzles were lifted without rotation, and the technique was thus aimed at producing vertical panels of cemented material.

The CCP and Jet Grout methods came to the attention of European companies in general and Italian companies in particular on the occasion of the international competition on the methods for the stabilisation of the Pisa Tower in the early 1970s, when the use of CCP (proposed by Konoike Co., Ltd.) was one of the five solutions judged worthy of mention. Commercial agreements were then initiated between some Italian companies and the Nissan Freeze Company, the owner of the patent of the CCP method, and the penetration of the technology into Europe began.

A further development led to the Column Jet Grout (CJG), also known as the 'Kajima' method, from the name of the Japanese company that developed it. This method consisted of simultaneous lifting and rotation of the rod, the lower part of which included an upper nozzle from which water and compressed air were injected and a lower nozzle that ejected cement grout (Yahiro et al. 1975).

As a result of this historical evolution, an important distinctive feature of each jet grouting technique is the kind and the number of fluids injected into the ground. Nowadays, the available techniques can be grouped into three main systems, which are named single, double and triple fluid, depending on the number of fluids injected into the subsoil, namely, grout (usually water–cement mixture), air and grout, and water plus air and grout.

In the beginning, jet grouting was mostly viewed as a means of improving the subsoil properties for the foundations of large structures. In the 1980s, jet grouting became very popular in Italy (Garassino 1983; Aschieri et al. 1983; Balossi Restelli and Profeta 1985; Tornaghi and Perelli Cippo 1985), Germany (Bell 1983; Berg and Samol 1986) and the United Kingdom (Coomber and Wright 1984; Coomber 1985a,b). It was then generally accepted in Europe as a reliable geotechnical tool, and its application was diversified for use in foundations, excavations, tunnelling, water barriers and underpinning.

Jet grouting entered the US market in the early 1980s (e.g., Langbehn 1986) but had a 'sluggish start' (Tarricone 1994), mainly because of perceived legal risks connected to the unknown technology (Xanthakos et al. 1994). However, after some time, the technology gained popularity in the United States and, subsequently, in Canada because it provided practical and cost-effective solutions to a number of difficult situations, such as for excavation support, ground-water barrier, bottom sealing to prevent pollutants from entering excavations, protection of bridges against scour, stabilisation of slopes and underpinning of existing foundations in commercial and industrial settings.

During the same period, jet grouting also became popular in most countries of South America, particularly in Brazil (Guatteri et al. 1988). Currently, jet grouting is used all over the world (e.g., Fang et al. 1994a; Ryjeski et al. 2009).

1.3 THE REASONS FOR SUCCESS

At first glance, the long-lasting success of jet grouting may seem rather surprising when compared with other soil improvement techniques. However, this fact can be easily explained by considering the unique outcomes that can be achieved by jet grouting. These include the following:

- Creating large columns of cemented material by drilling small holes into the ground, with limited disturbance of the surrounding subsoil
- Assembling such columns to form continuous elements of various shapes and sizes, provided with good mechanical properties and very low permeability

In some specific cases (e.g., water sealing the bottom of open excavations in urban environments), jet grouting has offered new geotechnical solutions to practical problems, leading to a substantial improvement of the construction process and thus changing the design philosophy.

The jet-grouted elements can also be reinforced with the insertion of metal or fibreglass bars or tubes, which provide a relevant flexural and tensile resistance if needed. It is thus possible to solve most geotechnical problems, such as increasing bearing capacity and reducing settlements of new or existing foundations, supporting open and underground excavations and creating water cut-offs for hydraulic reservoirs or groundwater barriers for pollution control. The most typical applications are sketched in Figure 1.1. In many geotechnical projects, jet grouting has thus become an effective alternative to traditional geotechnical techniques such as piles.

Jet grouting is also attractive because soil treatment can be performed even in difficult operating conditions and with relatively light equipment, working in confined spaces or in places difficult to reach by other means. Therefore, it can be conveniently used not only for the construction of new facilities but also for protecting or remediating existing ones.

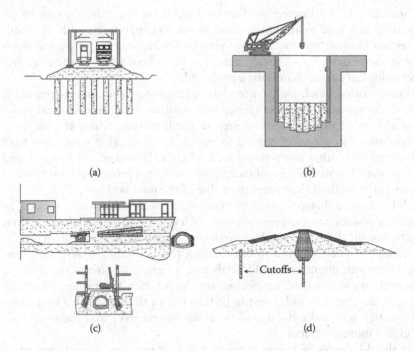

Figure 1.1 Some typical applications of jet grouting: (a) embankment foundations, (b) water sealing bottom plug and excavation support, (c) provisional tunnelling support and (d) water cutoff.

1.4 WHY A BOOK ON JET GROUTING?

Although jet grouting may seem a quick and easy way to solve most geotechnical problems, the use of this technology should be subjected to careful judgement. In fact, a thorough assessment of the literature reveals, at least in relative terms compared with other more conventional geotechnical techniques, the scarcity of detailed experimental studies on the effects of jet grouting, intended as the dimensions of the columns and their physical and mechanical properties. It follows that, as a rule, the characteristics of the jet-grouted columns are fixed by the designer more or less arbitrarily, and the treatment effects are then verified experimentally at the beginning of the work, based on the results of the so-called 'field trials'. Evidently, this process leads to a relevant uncertainty about the consistency between design assumptions and the actual characteristics of the jet-grouted elements, with obvious risks on matching the cost and time of work, which may lead to cumbersome litigation.

Sometimes, the limited confidence placed on the effectiveness of jet grouting may lead to an underestimation of its capabilities, with the result of either limiting its functions to provisional use only or inducing over-conservative design. As a result, more costly, and not always more effective, technological solutions are often preferred.

On the other hand, jet grouting has been adopted sometimes even when it did not represent the most appropriate solution to the geotechnical problem at hand. Examples are reported in the literature, where the inappropriate use of jet grouting has led to significant issues that would not have occurred with other more suited technologies. However, these errors and misuses stem more from ignorance about the actual possibilities and limitations of the method than from the value of the method itself.

A fundamental step toward overcoming these problems is to reduce, as much as possible, the inappropriate use of jet grouting and to adopt rational and scientifically based methodologies to design, execute and manage the technique. Observing and understanding the mechanisms activated during treatment, quantifying and predicting jet grouting effects and defects, knowing its limitations, foreseeing and monitoring the behaviour of the jet-grouted elements and, eventually, framing all this knowledge into codes of practice will make the use of jet grouting less risky and the practising engineer more confident.

It should always be borne in mind that a jet grouting project presents a number of specific problems that have to be considered both at the design stage and during construction. The assessment of jet grouting performance also requires specific measuring techniques and careful data processing.

To conveniently manage these aspects, it is important to have detailed knowledge of the various technological procedures that could be used and, above all, to understand the possible effects of jet grouting treatments on different natural soils. In recent years, this latter aspect has been subjected to thorough studies, composed of both physical and numerical modelling, which have substantially increased the knowledge of the complex jet–soil interaction phenomena. The results of these studies, supported by careful *in situ* investigations, have recently provided enhanced and reliable methods to estimate the effects of jet grouting, based on soil properties and treatment procedures. However, such studies are still mostly confined to the scientific literature, known only to a small academic community, and are not yet widespread among practising engineers.

Another crucial issue is the analysis of jet-grouted structures. This topic is of paramount importance for providing rational design methods that are urgently needed for establishing jet grouting as a fully reliable and mature technology. Again, rational solutions are available nowadays yet they are not known to most practising engineers who still tend to rely on purely empirical rules of thumb, mostly proceeding on the basis of personal experience and using some sort of trial-and-error procedure.

In writing this book, the authors have combined their personal experience on jet grouting projects and research with the available collective knowledge on jet grouting as reported by the technical and scientific literature, with the aim of providing an overall review of jet grouting, which could be useful to the technical community.

In fact, most of the educational publications on jet grouting tend to concentrate on the newest innovations and on successful applications of the technique, being rarely focused on highlighting and understanding defective behaviour or on developing effective design rules.

The lack of a commonly acknowledged design procedure is reflected in the existing National Rules, Codes of Practice and Guidelines on jet grouting. In fact, the guidance provided in such codes is usually meagre and not consistent worldwide.

Bearing in mind these considerations, this book presents an overview of recent research activities on jet grouting, with the perhaps ambitious aim of putting them into a unique, interconnected and consistent framework. The book, therefore, tries to tie together, in a rational way, the technological issues, the interpretation of the mechanisms occurring during jet grouting, the quantitative prediction of their effects, the design of jet-grouted structures and, finally, the procedures for controlling the quality of the constructed product (either a single column or a more complex jet-grouted structure).

Following this logical subdivision, the chapters of the book cover each of the aforementioned aspects. In particular, Chapter 2 describes the technological procedures, providing the details of the different jet grouting systems with their most recent developments. Chapter 3 focuses on the interaction between jets and soils, by reviewing the experimental investigations of the mechanisms occurring during the diffusion of jets and at their impact with the soil. Chapter 4 reports on the expected outcome of treatment, that is, the geometrical characteristics of columns and the mechanical and hydraulic properties of jet-grouted material, and provides some practical methods to calculate the effects of jet grouting in different types of soil. Chapter 5 illustrates the most typical applications of jet grouting, focusing on the functions of the jet-grouted elements, on the different possible geometrical arrangements of columns and on their possible defective behaviour. Chapter 6 discusses the principles of jet grouting design, proposing different alternative approaches for managing the variability of the geometrical and mechanical properties of the jet-grouted columns. Such design principles are then directly applied in Chapter 7, considering simple design schemes that provide practical calculation examples for the most frequent jet grouting applications. Finally, Chapter 8 presents the available experimental procedures for controlling jet grouting projects by considering the quality of materials, the treatment procedures, the performance of jet-grouted elements and structures, as well as the effects on the surrounding structures.

Chapter 2

Technology

2.1 TECHNOLOGICAL FEATURES

The jet grouting technology is based on the high-velocity injection of one or more fluids (grout, air, water) into the subsoil. The fluids are injected through small-diameter nozzles placed on a pipe that, in its usual application, is first drilled into the soil and is then raised towards the ground surface during jetting. Alternative jetting procedures have been developed in time and will be presented in this chapter. In most cases, the jets propagate orthogonally to the drilling axis, inducing a complex mechanical phenomenon of soil remoulding and permeation, with partial soil removal.

The injected water-cement (W-C) grout cures underground, eventually producing a body made of cemented soil. Most of the time, the treated volume has a quasi-cylindrical shape and is thus named 'jet-grouted column' or simply 'jet column'. There is, however, the possibility to make cemented bodies of different shapes, either by changing the treatment procedure or by joining several partly overlapped columns.

From the early stages of jet grouting, a variety of technical solutions has been proposed to increase the dimensions of the jet-grouted elements and to obtain columns as regular and homogeneous as possible. In fact, nowadays, there are several different ways to perform jet grouting, and new alternative procedures are frequently proposed. Many of them are patented, and some different commercial names are used. Therefore, a detailed description of all available jet grouting methods would not be practically feasible and would most likely soon become outdated.

The main technological features of the different procedures are, however, described in this chapter, considering the various treatment procedures, the most relevant equipment and the materials needed for jet grouting.

2.2 JET GROUTING PROCEDURE

Jet grouting is accomplished through the so-called 'jet grouting string'. The string is made by jointed rods provided with single, double or triple inner conduits, that convey the fluids to a tool, named 'monitor', mounted at the end of the string. The monitor is provided with one or more small-diameter nozzles, designed to transform the high-pressure fluid flow in the string into high-speed jets.

Different procedures can be chosen by selecting a proper combination of drilling and grouting. Usually, both operations are performed by using the same rig, which is able to regulate the rotation and translation of the jet grouting string and monitor (Figure 2.1).

Drilling is executed, up to the maximum desired depth of treatment, by using a rotating or rotary-percussive direct drilling system. The bit is mounted at the tip of the monitor and is slightly larger than the pipe string, thus leaving an annular space between the pipe and the borehole wall. The borehole diameter usually ranges between 120 and 150 mm, but, in some cases (see Section 2.3), it may be as large as 300 mm. Drilling can be performed with air, water, grouts or foams as flushing media. In general, the direct circulation of the drilling fluid, which flows downhole inside the hollow rods and uphole along the outer annular space, allows carrying of the drill cuttings to the surface and may also help in stabilising the borehole walls.

Jetting is then performed through one or more nozzles placed on the monitor, which is rotated and raised back. The soil is remoulded by the jet action, and part of the injected fluids and of the soil rise to the surface through the gap between the pipe string and the borehole wall, forming a sort of mud that is usually called 'spoil'.

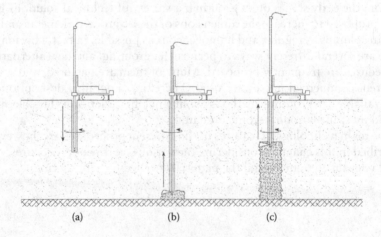

(a)   (b)   (c)

Figure 2.1 Typical jet grouting procedure: (a) drilling; (b and c) jet column formation.

Raising is usually carried out intermittently, with subsequent steps of 40 to 100 mm, whereas, for each lifting step, a number of rotations are performed (Figure 2.2a). In some cases, raising is performed continuously by retrieving the pipe at a constant rate, and thus, each jet moves along a spiral path (Figure 2.2b). In general, however, intermittent lifting is largely preferred because it makes treatment more effective, allowing more than one pass in the same points.

Whatever lifting procedure is chosen, the rotational speed is usually kept constant, and thus, a cylindrical body of cemented soil is finally obtained (Figure 2.3a).

If, however, the grout pipe is withdrawn without rotation, a thinner element, named 'jet-grouted panel', can be obtained. In particular, a planar element (Figure 2.3b) will be obtained if the nozzles are placed on the opposite sides of the monitor, whereas a V-shape one (Figure 2.3c) will be the result of treatment if the nozzle axes are set to form an angle on the horizontal plane.

More complex shapes may also be obtained by limiting the rotation angle and by rotating the pipe alternatively clockwise and counterclockwise or by changing the rotational speed during jetting. In particular, by placing the nozzles one opposite to the other and by varying the rotational speed along each turn, a candy or butterfly shape can be achieved (Figures 2.3d and 2.4). In principle, other polygonal shapes may be obtained by varying the rotational velocity (Shibazaki 2003), but at the moment, this application seems too complex to be implemented for routine jet grouting production.

In any case, the vast majority of jet grouting applications makes use of a proper combination of cylindrical elements, and thus, this book will deal almost exclusively with jet-grouted columns, either separated or interconnected by partial overlapping.

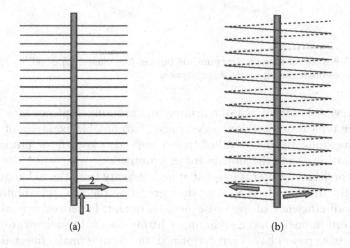

(a)

(b)

Figure 2.2 Rig lifting methods: (a) intermittent (1, lift; 2, jetting); (b) continuous (spiral path).

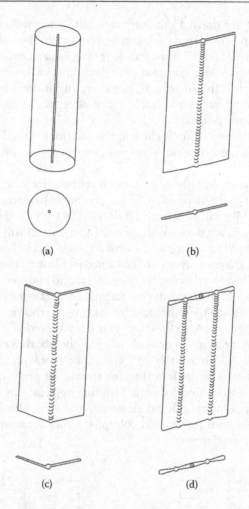

Figure 2.3 Possible shapes of cemented soil bodies: (a) column, (b) panel, (c) V-shape element and (d) candy-shape element.

As previously mentioned, the drilling and grouting sequence may differ from the typical one previously described. Two notable examples of alternative methods are the so-called 'prejetting' (also known as 'precutting' or 'prewashing') and the 'downward jet grouting'.

With prejetting, water is injected at high velocity from the radial nozzles during the drilling phase to start disaggregating the soil, thus improving the overall efficiency of the subsequent treatment. Prejetting may also be carried out in one or more preliminary lifting stages. In such a case, after the prejetting passes have been performed, the grout is finally injected from the bottom during a final lifting stage. In this particular case, the overall

Figure 2.4 Example of candy columns. (Courtesy of Trevi.)

procedure is often called a 'double treatment system'. Whatever way it is done, prejetting may be a cost-effective way to improve the treatment radius in those soils that are more resistant to erosion, such as dense sands or stiff clays.

The downward jet grouting procedure consists of performing the radial high-speed grout injection during the drilling phase. This method may also be combined with prejetting: in such a case, prejetting is carried out to a certain depth, and then downward treatment is performed (Sanella 2007). The depth (typically between 1/3 and 1/2 of the total column length) at which the two phases (prejetting and downward treatment) are switched must be tuned case by case to minimise the spoil outflow. With this aim, the spoil characteristics are continuously monitored at the ground surface to check its composition and to control the effectiveness of the treatment procedure. In the first prejetting phase, the spoil is just a mixture of water and removed soil, having a low density. A sharp increase in the spoil density indicates that the upward flow of the injected grout has been able to reach ground level. The experimental evidence indicates that, when this happens, the column has been formed up to ground level and thus also above the depth from which downward jet grouting was started. With this approach, a limited amount of spoil may be produced, and cost-effective production may also be achieved.

2.3 JET GROUTING SYSTEMS

The most important distinctive feature of each jet grouting technique consists of the kind and number of fluids injected into the ground (Yahiro and Yoshida 1973). Typically, the available techniques are grouped in three main jet grouting systems, named single, double and triple fluid systems, depending on the number of fluids injected into the subsoil. The fluids are grout (usually a W-C slurry) for the single fluid; air and grout for the double fluid; and water, air and grout for the triple fluid (Figure 2.5).

According to the number of fluids, the jet grouting strings are made of different pipe types, and the monitor is provided with different features (Figures 2.6 and 2.7). The monitor is a steel cylinder placed at the end of the jet grouting string, immediately above the cutting tool. It houses the nozzles from which grout slurry, air and water are injected during treatment, and as a consequence, it is the key tool of jet grouting. Typically, the nozzles from which the cutting fluid is ejected have diameters within the range of 2 to 8 mm. For reasons linked to the hydrodynamic efficiency of jet grouting (see Chapter 3), the tendency nowadays is to reduce the number of nozzles, adopting diameters close to the upper boundary of the aforementioned range.

The monitor has a hole at the bottom, much larger than the nozzles, which is used during the drilling phase for the direct circulation of the drilling fluid. When the jetting phase starts, the bottom hole is closed, and the high-pressure fluids are injected via the lateral nozzles.

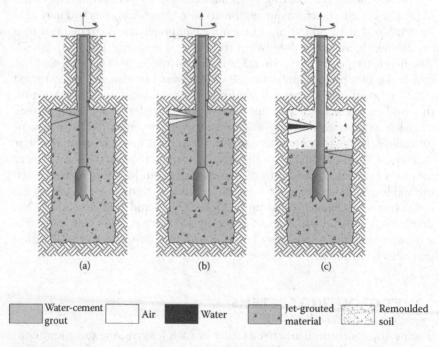

| Water-cement grout | Air | Water | Jet-grouted material | Remoulded soil |

Figure 2.5 Typical jet grouting systems: (a) single fluid, (b) double fluid and (c) triple fluid.

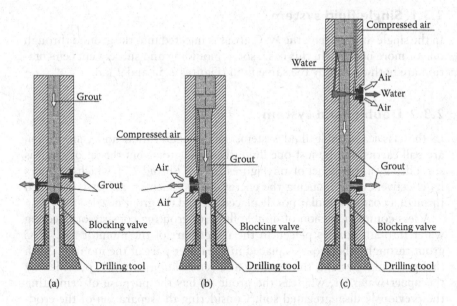

Figure 2.6 Schematic drawings of the monitors: (a) single fluid, (b) double fluid and
(c) triple fluid.

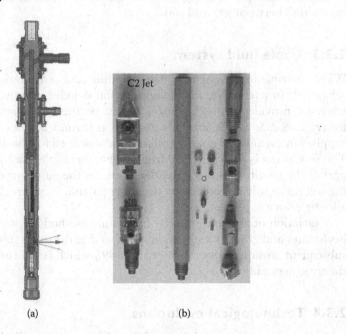

Figure 2.7 Example of a commercial double fluid jet grouting equipment: (a) schematic
section of the drilling and injection string and (b) photograph of the components. (Courtesy of CRM)

2.3.1 Single fluid system

In the single fluid system, the W-C grout is injected into the ground through one or more nozzles. In this case, soil remoulding and subsequent cementation are both caused by the same fluid (Figures 2.5a and 2.6a).

2.3.2 Double fluid system

In the classical double fluid system, soil disaggregation and cementation are still carried out by just one fluid, the W-C grout, but the jet of grout is shrouded by a coaxial jet of air (Figures 2.5b, 2.6b and 2.7), which enhances its effectiveness by reducing the energy losses. Such an air jet is provided through a coaxial annular nozzle placed around the grout nozzle.

A less common version of double fluid jet grouting consists of injecting water through nozzles placed in the upper part of the monitor and W-C grout through other nozzles placed in the lower part of the monitor. With this particular system, the eroding and remoulding action is provided by the upper water jet, whereas the grout jet has the purpose of cementing the previously disaggregated soil. Considering the separation of the eroding and cementing actions, this latter version of the double fluid system resembles the triple fluid jet grouting system in terms of the mechanical interaction between jets and soil.

2.3.3 Triple fluid system

With the triple fluid system, soil remoulding and cementation are clearly separated. In particular, soil disaggregation is induced by a high-velocity water jet, provided through a nozzle placed on the upper part of the monitor (Figures 2.5c and 2.6c). This water jet is shrouded by a coaxial air jet, supplied by an annular nozzle similar to the one used for the double system. The W-C grout is then delivered from a separate nozzle placed on the lower part of the monitor. In this case, the grout has the only purpose to cement the soil previously remoulded by the water jet and, therefore, is delivered at a lower velocity.

A variation of the conventional triple fluid method consists of injecting both water and grout at a very high speed so that the soil is subjected to two subsequent erosion stages (Shen et al. 2009), which could further enhance the treatment radius.

2.3.4 Technological evolutions

In general, the erosive capability and thus the treatment radius progressively increases, moving from a single to a double and, then, to a triple system. Whatever the jet grouting system, the nozzles should be carefully designed for reducing, as much as possible, the localised head losses to deliver a jet with the highest possible velocity. Therefore, the nozzle shape plays a relevant role, as discussed in detail in Chapter 3. A technological detail that deserves to be anticipated is that jet efficiency increases by making, as long as possible, the rectilinear delivering stretch that constitutes the outer part of the nozzle and also by minimising the curvature of all the curved parts of the fluid conduit (Shibazaki 2003; Shibazaki et al. 2003).

The manufacturing companies have been working for several years and still keep on working on this and other important details to improve the overall performance of jet grouting.

An example of the numerous attempts to improve the effectiveness working on nozzles is the so-called 'cross jet method' (Shibazaki et al. 1996; Shibazaki 2003). With this method, the monitor houses two nozzles, one on top of the other, oriented to produce inclined jets, colliding at a given distance from the monitor axis (Figure 2.8). According to Shibazaki (2003), more regular columns can be obtained by tuning such a distance. However, this method has not found a wide application so far.

In recent years, because of the technological progresses, much more powerful pumps and more efficient supply circuits have been adopted in jet grouting, allowing to reach extremely large diameters of the columns. Very well-known and, nowadays, relatively popular examples are the so-called 'SuperJet' and 'SuperJet Midi' (Yoshida et al. 1996). With SuperJet, columns having a diameter of up to 5 m can be obtained. The distinction among the two commercial names is related to the different flow rates and

Figure 2.8 Simplified scheme of the cross jet method: the two inclined jets collide at a desired distance from the monitor.

to some details of the monitor. The improvement in jet grouting capacity is, in this case, related to a much more efficient design of the supply circuit and of the nozzles, with a much larger hydrodynamic efficiency that minimises flux turbulence and hydraulic head losses (see Chapter 3).

However, with this technique, the monitor cannot house a cutting tool and, as a consequence, it must be placed in a previously drilled borehole, possibly sustained by bentonitic or polymeric slurry or by temporary casing. Furthermore, the borehole is much larger than usual (typically 300 mm) to create a larger annulus clearance around the string pipe, which is necessary to guarantee the continuous spill of the larger amount of spoil.

2.4 COLUMNS OVERLAPPING

Many applications of jet grouting are based on the creation of jet-grouted elements of various customised shapes, obtained by partly overlapping a number of jet-grouted columns. To this aim, two alternative production sequences may be adopted, named 'fresh in fresh' and 'fresh in hard'.

With the fresh-in-fresh sequence (Figure 2.9a), the adjacent columns are created by performing the contiguous treatments at short time intervals, without waiting for grout hardening. In this way, part of the previously created column can be eroded and injected again by the new ongoing treatment, and thus, a continuous cemented volume can be obtained. However, particular care must be placed during the drilling phase to avoid the washout of the previously created and still-fresh columns by the drilling water.

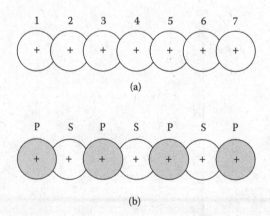

Figure 2.9 Fresh-in-fresh sequence (a) and fresh-in-hard sequence (b) for a rectilinear jet-grouted element (P and S, respectively, for primary and secondary columns, with reference to the production sequence).

In the most critical situations, this objective can be met by using W-C grout as drilling fluid.

With the fresh-in-hard sequence (Figure 2.9b), also called the 'primary-secondary sequence', each new column is created only after waiting for a relevant hardening of the adjacent columns. If this sequence is used, it is necessary to make a detailed treatment plan composed of primary, secondary and, sometimes, even tertiary column sets. Such a treatment plan should be properly specified, depending on the kind of structure to be formed (Figures 2.10 and 2.11). The fresh-in-hard sequence is the most popular one but poses the problem of the so-called 'shadow effect' because the previously hardened columns do not allow the jet to reach and erode the soil on the sides of such columns, thus reducing the dimensions of the secondary and tertiary columns (van Tol 2004).

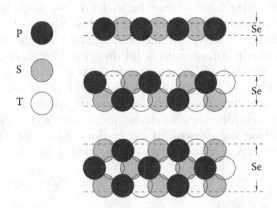

Figure 2.10 Examples of fresh-in-hard sequence of groups of overlapped columns (P, S and T, respectively, for primary, secondary and tertiary columns).

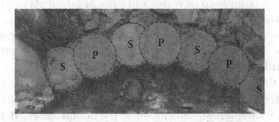

Figure 2.11 Photograph of a curved horizontal element for a tunnel canopy made by primary (P) and secondary (S) columns. (From Mandolini, A., Manassero, V., Interventi Geotecnici di Carattere Strutturale, Proceedings of the 24th National Conference of Geotechnical Engineering 'Innovazione tecnologica nell'Ingegneria Geotecnica', Napoli (Italy), pp. 5–49, 2011 [in Italian].)

2.5 COLUMNS REINFORCEMENT

Single jet columns and larger jet-grouted elements made by multiple columns can be reinforced by inserting steel bars, tubes or other reinforcing elements to provide some tensile and flexural strength to the overall structure. Steel reinforcements may be directly pushed into the fresh column or inserted after the total or partial hardening of the grout by drilling a hole into the column and sealing the reinforcement with conventional low-pressure grouting. The previous option is feasible only for relatively shallow and short jet grouting because pushing the reinforcement into the freshly made column becomes extremely difficult at depth. On the contrary, drilling and grouting can be carried out at any depth but makes the overall production process more complex. In all cases, particular care is needed to place the reinforcements in the desired position.

In some particular cases, it may be useful to use fibreglass elements made of a polymeric matrix reinforced by glass fibres. The glass fibres are oriented along the axis, and thus, the mechanical behaviour of these elements is markedly anisotropic. It follows that fibreglass elements are very flexible, provided with almost no flexural resistance but high tensile strength, ranging between 600 MPa for the tubes and 1000 MPa for the bars. Such fibreglass reinforcements are particularly suitable for the provisional reinforcement of tunnel face because they can be easily demolished during the subsequent excavation phases. Because of their flexibility, fibreglass reinforcements can be placed only by drilling and grouting.

2.6 GROUT MIX AND SPOIL

2.6.1 Grout mix

The grout mix is composed of water and cement dosed according to weight ratios, W/C, usually ranging between 0.6 and 1.3. The most appropriate W/C ratio should be chosen, for each specific project, considering that increasing the W/C ratio will result in a higher erosion efficiency (see Chapter 3) but lower strength of the jet-grouted material.

In general, there are no particular restrictions on the type of cement to be used, although its characteristics have to be chosen considering the site conditions (see Chapter 8), and in some particular cases, it may be useful to choose special cement types. For example, if rapid setting times are required, it may be convenient to use fine Portland cement, whereas, in a chemically aggressive environment, the use of pozzolanic or blast-furnace cement may be required.

The addition of additives may also be useful for some applications. The most widely used is bentonite, which can be added in the form of suspension to reduce bleeding, especially when the W/C ratio of the grout is very high and/or high strength is not an issue.

Other additives that, in some cases, may be used are calcium chloride, to accelerate grout hardening, and sodium silicate, to accelerate grout setting and thus to prevent possible washout by underground water flow. The latter must be handled with extreme care because it can make setting almost immediate. Usually, there are two ways to add it to the grout: the first is by inserting a static mixer after the agitator, immediately before the high-pressure pump, to reduce the risk of gel setting in the circuit; the second is by injecting the sodium silicate through a separate conduit so as to mix it with the cement directly in the ground.

2.6.2 Spoil

As previously mentioned, during jet grouting, part of the injected grout and of the eroded soil will rise to the surface through the annular gap formed between the pipe string and the borehole wall, forming the spoil (Figure 2.12). Clearly, the spoil is a waste that, in principle, should be minimised for cost effectiveness and for reducing the problems posed by its disposal or recycling under the ever more restrictive environmental rules.

(a)

(b)

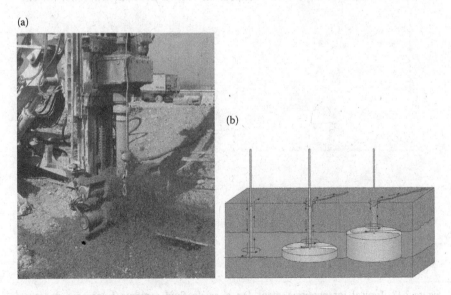

Figure 2.12 Jet grouting spoil outflow (a) and production (b).

However, a moderate amount of spoil is essential for gaining an effective treatment because the upward flow of the spoil along the borehole guarantees that there is no hole clogging. If, on the contrary, the borehole collapses or, for any other reason, there is no free connection between the annular space around the monitor and the ground surface, the grout pressure may rapidly build up, eventually fracturing the soil once the pressure overcomes soil resistance. In such an undesired circumstance, instead of creating a jet column, the treatment may result in thin grout layers. This may also cause heave at the ground surface and possible dangerous movements of the nearby existing structures.

Therefore, in case the collapse of the borehole is feared, some ways to sustain the borehole walls should be considered. A direct circulation of bentonite, cement or polymeric slurry may be used in the drilling phase. A casing may be adopted as well, although it makes jet grouting production much more cumbersome and expensive because casing extraction and jetting phases should usually be alternated.

2.7 JET GROUTING PLANT

2.7.1 Fundamental tools

A typical jet grouting plant is organised according to the scheme reported in Figure 2.13, which refers to the simplest case of a single fluid system. For the double fluid system, an air compressor is also needed, whereas, for the

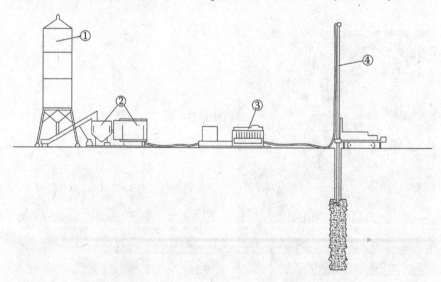

Figure 2.13 Typical jet grouting plant for a single fluid system (1, cement storage; 2, dosage and mixing plant; 3, high-pressure pump; and 4, drilling equipment).

triple fluid system, an air compressor and a high-pressure water pump are both required.

The fundamental tools are, therefore, the manufacturing system of the W-C grout, the grout and water pumps, the air compressor, the drilling rig, the string rods and the monitor and the hydraulic circuits (high-pressure hoses and connecting junctions).

2.7.2 Grout mixing

The W-C grout is manufactured by means of a preferably automatic plant, which must ensure a continuous production, with possible registration of the number of mixes and of the quantities of the individual components for each single mix. The cement content is determined by a scale; the water and the possible additives are regulated by another balance or by a volumetric dosing device.

Mixing occurs in a high-turbulence primary mixer. At the end of the primary mixing cycle, the mixture is conveyed to a storage container provided with a low-speed agitator and then to the pump. The grout mix production required for each workstation is generally between 10 and 20 m³/h.

2.7.3 Pumps and compressors

The key element of the jet grouting technology is the pumping system. In fact, very high-pressure pumps (with pressures of up to 50 MPa or even more) are required for pumping the grout, in the single and double fluid systems, as well as for pumping the water in the triple fluid system. To obtain such high pressures, special piston pumps are used, driven by a diesel engine and equipped with gears, to adjust the flow rates as requested. Such high pressures are not usually required for the grout if the triple fluid system is used because, in this case, the erosive action is demanded to the water jet. Therefore, the grout pressure for the triple fluid system does not usually exceed the value of 10 MPa. In some cases, lower pressures are used also with single and double fluid jet grouting to adjust for specific projects.

The air compressor, required in the double and triple systems, should be able to guarantee a pressure ranging between 1.2 and 2.5 MPa, with air flow rates in the order of 200–300 l/s.

2.7.4 Drilling and grouting rigs

The drilling and/or grouting rigs have different characteristics for outdoor and for underground works. Each grouting rig is equipped with a battery of hollow rods, with a diameter ranging between 60 and 140 mm. The rods used in the single fluid system have just one conduit and are thus stronger than the rods used in the double and triple systems, which host, respectively, two and three inner conduits. There are also special rods allowing rotary-percussion drilling, which may be needed to pass through very stiff soil layers, rock blocks or masonry.

The monitor and the drilling tool are mounted at the tip of the rig. The pipe junction is mounted at the top and connected to the pump and to the compressor through a number of hoses equal to the number of fluids.

2.8 TREATMENT PARAMETERS

Each jet grouting treatment is performed by regulating and controlling a set of technical features that can be called 'basic treatment parameters' and can be grouped as follows (Table 2.1):

- Geometrical characteristics of the mechanical device
- Kinematic variables defining the movement of the jet grouting string
- Composition, pressure and flow rate of the injected fluids

Clearly, each jet grouting system (single, double and triple fluid) will have a different set of parameters regarding the injected fluids. Some parameters (e.g., fluid pressures, nozzle diameter and flow rates) cannot be chosen independently because they are correlated.

In practice, reference is often made to treatment parameters derived from the basic ones listed in Table 2.1. The derived parameters are as follows:

- Average lifting speed of the monitor
- Monitor rotations for each lifting step
- Injected grout volume per treatment unit length
- Mass of injected cement per treatment unit length

Table 2.1 List of fundamental treatment parameters

Parameter	Definition	Symbol	Unit	
			S.I. unit	Practical unit
Geometrical	Number of nozzles	M	–	–
	Nozzle diameter	d	m	mm
	Lifting step	Δs	m	cm
Kinematic	Time interval per step	Δt	s	s
	Rotational velocity	ω	rad/s	round/min
	W-C ratio by weight	W/C	–	–
Injected fluids	Fluid pressure[a]	p_g, p_w, p_a	MPa	bar
	Fluid flow rate[a]	Q_g, Q_w, Q_a	m^3/s	L/min

[a] The pedex indicates, respectively, grout (g), water (w) and air (a).

Table 2.2 List of derived treatment parameters

Derived parameter	Relation with the parameters defined in Table 2.1	Unit
Average lifting speed of the monitor	$$v_r = \frac{\Delta s}{\Delta t}$$	mm/s
Monitor rotations for each lifting step[a]	$$n_g = \omega \cdot \Delta t$$	–
Injected grout volume per treatment unit length	$$V_g = \frac{Q_g}{v_r}$$	m³/m
Mass of injected cement per treatment unit length[b]	$$W_c = \frac{\rho_g \cdot V_g}{1 + W/C}$$	kg/m

[a] ω expressed in RPM.
[b] ρ_g is the grout density.

The relationship between basic and derived parameters is provided in Table 2.2. These derived parameters can be very useful in comparing different treatment procedures as well as for implementing empirical or numerical methods for estimating the jet column diameter.

Typical ranges of treatment parameters are finally provided in Table 2.3. They are only indicative because it may happen that, in some specific cases, values external to such ranges may be chosen. Furthermore, technological evolution in pumping apparatuses, monitors, nozzles and procedures is very fast, and the ranges reported in the table will probably be subjected to substantial changes in the near future.

Table 2.3 Typical values of treatment parameters

Treatment parameter	Symbol	Unit	System		
			Single fluid	Double fluid	Triple fluid
Lifting step	Δs	mm	40–50	40–80	40–100
Average lifting speed	v_r	mm/s	4–10	1–8	0.5–5
Rotational velocity	ω	rpm	5–40	3–30	1–40
Nozzle diameter	d	mm	2–8.0	2–8	2–8
Number of nozzles	M	–	1–2	1–2	1–2
Grout pressure	p_g	MPa	30–55	20–40	2–10
Air pressure	p_a	MPa	–	0.5–2.0	0.5–2.0
Water pressure	p_w	MPa	–	–	20–55
Grout flow rate	Q_g	L/s	2–10	2–10	2.0–5
Air flow rate	Q_a	L/s	–	200–300	200–300
Water flow rate	Q_w	L/s	–	–	0.5–2.5
W-C ratio by weight	W/C	–	0.60–1.25	0.60–1.25	0.40–1.0

Chapter 3

Mechanisms and effects

3.1 PRELIMINARY CONSIDERATIONS

High- and ultrahigh-speed water jets have been and are still currently used in industry for various purposes. Table 3.1 (Summers 1995) reports the pressures and flow rates for some of the most common applications, such as rock cutting (Farmer and Attewell 1965) and metal cleaning (Conn and Gracey 1988; Momber and Kovacevic 1997).

The jetting technique was originally conceived for open-air use, and only in a second moment its use was extended to other more complex boundary conditions to accommodate particular needs. In civil engineering, for instance, underwater applications, such as cutting of marine structures or drilling at the sea bottom, have been reported (Iwasaki 1989).

Experimental investigations carried out by processing high-resolution images taken with high sampling rate cameras or by detecting the change of colour on pressure-sensitive films (e.g., Soyama et al. 1996) have revealed that high-velocity jets are characterised at their boundary by a complex sequence of vortex cavitation mechanisms, whose analysis and understanding has been essential to conceive more and more efficient systems. In industrial applications, these basic hydrodynamic studies helped, several decades ago, in finding the best shape and diameter of the ejecting nozzle and of the feeding circuit, as well as the most effective distance between the nozzle and the target.

In the mid-1960s, these experiences and their industrial applications were the basis of the idea of jet grouting. The first experiments were carried out with a chemical grout, but soon this binder was replaced by cement grout because of economic and environmental reasons.

In a jet grouting process, the cutting fluid emitted from the nozzles (grout or water, depending on the adopted system) has to cross the annular space formed between the rod and the undisturbed soil to erode and permeate the ground. Since soil erosion can be satisfactorily achieved only if the fluid velocity at the impact with the undisturbed soil is high enough, a fundamental role in the effectiveness of treatments is played by the diffusion of

Table 3.1 Examples of water jet applications

Application	Operational pressure (bar)	Flow rate (L/min)
Car wash and cleaning	70	20
Coal and rock mining	70	4000
Industrial cleaning	140–1400	20
Mining and demolition	700–1000	40–200
Industrial machining	2000–4000	4
Impulse fragmentation	2000–7000	40–80
Special application	>70,000	Varied flow

Source: Summers, D. A., *Waterjetting Technology*: London, United Kingdom: E & FN Spon, 882 p., 1995.

the jet within the composite fluid material filling this gap, made of groundwater, remoulded soil particles and previously injected grout.

Bearing in mind this goal, noticeable efforts have been spent from jet grouting pioneers in developing hydrodynamic systems as efficient as possible to preserve the cutting capacity of the jet at the largest possible distance from the nozzle.

Although the jet grouting technique was initially conceived on an empirical basis (Shibazaki et al. 1983; Yahiro et al. 1983; Miki and Nakanishi 1984), the need for capitalising on the pioneering experimental observations into theories describing the diffusion of submerged jets soon emerged. Recent developments in the techniques, the procedures and the models available to predict the result of treatment have demonstrated that significant progress has been achieved. Very large columns can be now created, and predictions can be made with reasonable reliability (see Chapter 4).

As will be shown in some detail in the following, the analysis of jet grouting has to consider two subsequent mechanisms:

- The propagation of the jet across the space included between the nozzles and the undisturbed soil
- The interaction between the jet and the soil

At the beginning of treatment, the distance between the nozzle and the intact soil is relatively small, being created by an excavation toe only slightly larger than the monitor. Thereafter, as far as erosion occurs, the soil boundary shifts outwards, and the fluid region becomes larger.

After reaching the undisturbed soil, part of the injected fluid (q_1 in Figure 3.1) keeps the radial direction (radial flow) eventually displacing the soil grains, whereas another non-negligible part (q_2 in Figure 3.1) flows toward the ground surface through the annular space around the injection stem, carrying along a part of the eroded soil. The spoil flowing to the surface may assume largely variable values (between 0% and 80% of the

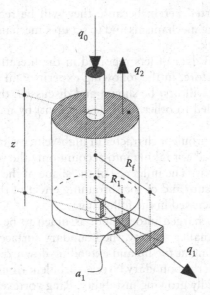

Figure 3.1 Radial flow and spoil return in jet grouting. (From Modoni, G. et al., *Géotechnique*, 56, 5: 335–347, 2006.)

injected flow, according to Kaushinger et al. 1992). Usually, spoil outflow tends to decrease as soil grain size increases. In all cases, the outflow is not constant because the spoil comes out from the borehole at irregular intervals, with a variable flow rate.

Based on a combination of experimental observations and on some simple theoretical formulations, this chapter provides an interpretation of the mechanisms (diffusion of submerged jets and interaction with different soil types) leading to the formation of a jet grouting column. The influence of the most important technological details presented in Chapter 2 will also be considered. The considerations reported in this chapter provide the basis for the formulas reported in Chapter 4 to predict the average diameter of jet grouted columns.

3.2 SUBMERGED JET

3.2.1 Experimental observations

The complex physics of submerged jets has been extensively investigated both experimentally and theoretically by many researchers (e.g., Hinze 1975; Davidson 2004). Although a detailed analysis is beyond this book, some basic experimental and theoretical information on fluid dynamics are reported herein because they will be recalled and used to interpret jet grouting mechanisms and to set up a mechanically sound interpretation framework.

The experimental data of jets reported in the literature mostly refer to jets of water. Therefore, in the following experimental evidence on water jets into still water will first be shown and discussed; the obtained results will then be extended to other fluids (e.g., grout) by means of theoretical considerations.

Considering the turbulent character of high-velocity jets, their characterisation will be carried out taking into account only the velocity component parallel to the jet axis. The influence of the shape of the nozzle, of the surrounding fluid pressure and of the shrouding effect of the coaxial annular jet of air will be discussed in the following.

When a jet is discharged into a fluid, assumed to be quiescent because of its nil or much smaller velocity, the boundary surface of the jet forms a shear layer originating at the lip and extending downstream. In very high-speed jets, for which the boundary layer is turbulent from the nozzle, such a layer exhibits a rapidly growing instability. Ring vortexes formed in the jet shear layer are advected downstream, and they interact, merge and eventually become unstable, breaking down and increasing turbulence.

The associated evolution of the jet can be observed from the photograph reported in Figure 3.2a (Bergschneider 2002). Immediately after the exit from the nozzle, within a length of some nozzle diameters, the flow keeps a rather constant cross-section. In this initial region, the shear layer at the boundary of the jet does not affect the velocity v, which keeps values close to that of the nozzle outlet (v_0). In this initial part, the component of velocity orthogonal to the jet axis is negligible. At larger distances from the nozzle, the jet widens into a fan, increasing its cross-section. As shown in Figure 3.2, in this region, turbulent vortexes are observed within the fluid: the flow becomes unstable, exhibiting whirling waves and instantaneous turbulent components of velocity randomly oriented in the space. The jet turbulence induces the following two mechanisms to occur on its external boundary (Figure 3.2b):

- A mass exchange with the surrounding fluid, that increases the flow rate because part of the surrounding fluid is entrained into the jet
- A loss of energy caused by the vortexes and by the interface viscous shear stresses, with a reduction of the longitudinal component of the velocity

As a consequence of these two mechanisms, the jet widens into a fan with increasingly larger cross-sections. Efficient jets have to be as concentrated and fast as possible.

(a)

(b)

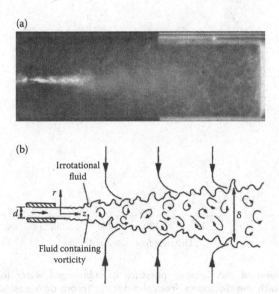

Figure 3.2 Photograph (a) and sketch (b) of a high-speed submerged jet. (From Bergschneider B., *Zur Reichtweite beim Düsenstrahlverfahren im Sand*, PhD thesis, Bergische Universität Wuppertal Fachbereich Bauingenieurwesen [in German], 2002.)

In each point of the flow field outside the nozzle, the velocity is randomly variable with time. It is convenient to consider two components for it: a mean one, obtained by integrating the velocity vector in a sufficiently long time interval, and a turbulent one having intensity and direction randomly variable with time. In the case of stationary boundary conditions, the mean component is a constant, whereas the mean value of the turbulent component is nil. For jet grouting, the mean longitudinal velocity component is the only one of interest.

Measuring the evolution of ultrahigh-speed submerged jets is not a simple task, basically because of the uncommon velocities involved in the process (in the order of some hundreds of metres per second) and the disturbance possibly induced to the flow by the insertion of measuring devices. For these reasons, the literature reports only a few examples of experimental investigations in which the velocities (or dynamic pressures) have been measured.

One of them (de Vleeschauwer and Maertens 2000) refers to a system based on the Pitot tube principle to record the variation of dynamic pressures along the axes of submerged water jets streaming from a nozzle having a diameter $d = 2.2$ mm. The experimental curves (Figure 3.3) obtained for different injection pressures ($15 < p_0 < 40$ MPa) show, for all cases, a

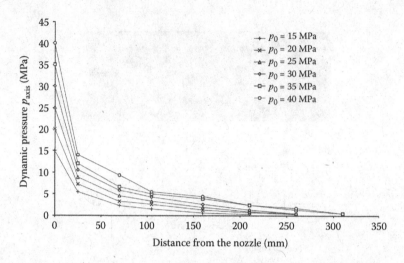

Figure 3.3 Evolution of the dynamic pressure of submerged water jets on the jet axis with the distance *x* from the nozzle. (From de Vleeschauwer, H. and Maertens, G., Jet grouting: State of the art in Belgium, *Proceedings of the Conference 'Grouting*–Soil improvement–Geosystem including reinforcement', Finnish Geotechnical Society, Helsinki: 145–156, 2000.)

sudden pressure drop immediately after the exit from the nozzle and a progressively smoother decay with distance.

This aspect can be better observed by plotting the same data in terms of the velocity decay v_{axis}/v_0 as a function of the normalised distance x/d from the nozzle, where v_{axis} represents the mean component of velocity oriented along the axis at a distance x from the nozzle, and v_0, the inlet velocity. Once the corresponding pressures are known, such a ratio can be calculated using the relationship $v_{axis}/v_0 = \sqrt{p_{axis}/p_o}$.

In Figure 3.4, the data from de Vleeschauwer and Maertens (2000) are compared with two longitudinal profiles presented by Di Natale and Greco (2000), who adopted the particle image velocimetry technique to study the evolution of jets flowing with relatively low velocities (0.5–1 m/s) from a nozzle having $d = 4$ mm. The most interesting effect shown in this plot is that the decay profile is not unique, depending on the initial velocity at the nozzle.

Reynolds number at the nozzle (Re = $\rho \cdot v_0 \cdot d_0/\mu$, where ρ and μ are, respectively, the density and the dynamic viscosity coefficient of the injected fluid, and d_0 is the diameter of the nozzle) can be conveniently used to summarise the observed effects. Larger Reynolds numbers imply larger hydrodynamic efficiencies. Because Re = $(3 \div 6) \times 10^5$ in the case of de Vleeschauwer and Maertens (2000), and Re = $(1 \div 2) \times 10^3$ in the experiments of Di Natale

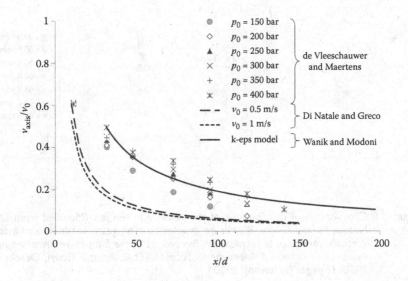

Figure 3.4 Velocity decay with the normalised distance x/d from the nozzle along the axes of submerged water jets.

and Greco (2000), the former experiments keep larger percentages of the initial velocity at a given normalised distance from the nozzle x/d. The formulation of Reynolds number also suggests that more efficient jets can be obtained by adopting denser fluids, as long as they do not increase the viscosity.

The velocity distribution in the different cross-sections of a jet has been observed in a number of experimental researches. As is clearly shown by the experiments reported by Di Natale and Greco (2000), the observed distribution of the velocity along the jet cross-section is bell shaped (Figure 3.5). Although these experiments have been conducted with low inlet velocities, it is reasonable to assume that, also in high-speed jets, the distribution of velocities in the cross-section is bell shaped. This experimental evidence confirms that energy dissipations at the flow boundary play a key role in reducing jet effectiveness, and therefore, the erosive efficiency of a jet can be improved not only by increasing the outlet velocity but also by trying to reduce the energy losses along the jet boundaries.

The kinematic characteristics of submerged jets also depend on the shape of the ejecting nozzle. For a given hydraulic head h at the nozzle, the outlet velocity v_0 can be calculated with the following expression:

$$v_0 = C \cdot \sqrt{2gh} \tag{3.1}$$

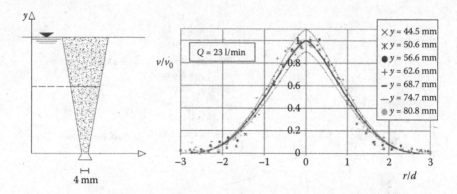

Figure 3.5 Cross-sectional distributions of the velocity in the jet. (Modified from Di Natale, M. and Greco, R., Misure di velocità in un getto sommerso ad asse verticale mediante la tecnica PIV, *Proceedings of the Symposium promoted by Associazione Italiana di Anemometria Laser (AIVELA)*, Ancona (Italy), October 2000: 11 pages [in Italian], 2000.)

in which g is the gravity acceleration, and C is a coefficient describing the energetic efficiency of the outflow, that, in turn, depends on the shape and dimensions of the nozzle.

A less intuitive result is described by the comparative experimental study presented by Leach and Walker (1966), who connected a pressure transducer to a steel plate hit by water jets flowing in air. Figure 3.6 shows the velocity decay profiles along the longitudinal axis obtained by moving the

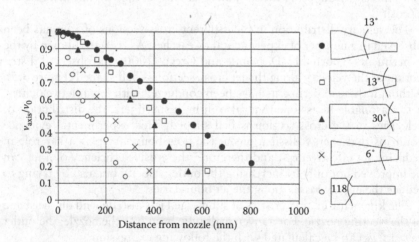

Figure 3.6 Decay of longitudinal velocity obtained for different nozzle shapes. (From Leach, S.J. and Walker, G.L., *Phil. Trans. Royal Soc. London. A*, 260, 1100: 295–308, 1966.)

steel plate at variable distances from the nozzle. The plot clearly demonstrates that the nozzle shape is also a key factor in determining the velocity decay. The best possible shape among the ones shown in the figure is the one having a smooth transition between two straight parts of the nozzle and a final straight part that is at least 2.5 mm long. In this nozzle, which is named 'Nickonov and Shavlovskj' after the Russian researchers who made hydrodynamic studies on it, the smooth transition minimises the concentrated head losses, whereas the straight final part ensures that the fluid threads at the exit are almost parallel, thus minimising the radial velocity component.

However, the shape of the nozzle itself is not the only detail of the monitor that plays a relevant role in keeping the jet velocity as high as possible. In fact, the nozzle's shape governs only the local reduction of velocity, whereas another key factor is the incoming velocity into the nozzle, which, for obvious reasons, must be taken as high as possible. This is indeed a critical technological detail because the nozzles are orthogonal to the feeding tubes located along the rig axis, and thus, the pressurised fluid is discharged perpendicularly to the incoming circuit. In the first versions of jet grouting monitors, such a change of direction of the grout or water flow inside the monitor circuit was abrupt, resulting into highly concentrated head losses and an extremely turbulent, dispersed and, therefore, ineffective jet. This is a very well-known problem in high-speed hydraulic circuits, and continuous evolutions of monitor design have brought to a significant reduction of concentrated head losses in the circuit, providing large improvements in efficiency. The most well-known example is related to the jet grouting system commercially known as 'Super Jet' or 'Super Jet Midi', depending on the adopted flow rate and on some details of the monitor, in which the hydraulic circuits have been carefully designed to take as large as possible the radius of all the curves.

In particular, Shibazaki et al. (2003) and Yoshida (2012) have shown that improving the shape of the circuit and of the nozzle significantly contributes to increasing the efficiency of jet grouting, reducing the energy needed at the pump to develop a desired flow rate. As a result, Yoshida (2012) claims that the CO_2 emission caused in Japan by jet grouting, which is an index of the energy cost, has reduced to one sixth compared with that calculated 20 years before. In their article, Shibazaki et al. (2003) show some very interesting laboratory results (Table 3.2) conducted to evaluate the effect of feed pipe diameter curvature on the jet energy. Using a water jet pressurised at 30 MPa, with a flow rate of 300 l/min, and measuring the jet pressure on a steel plate located at a distance of 5.5 cm from the outlet section, they found that the larger the rectification coefficient (defined as the ratio between the radius of the curve and the diameter of the pipe), the more concentrated and energetic the jet. For a given diameter of the pipe of 19.4 mm (first three lines of the table), for instance, when the coefficient of rectification changes from 2.06 to 4.90, the jet pressure doubles its value

Table 3.2 Effect of pipe curvature and diameter on jet energy

Internal diameter of the bent pipe (d; mm)	Radius of the curve (R; mm)	Rectification coefficient (R/d)	Widening angle of the outlet jet	Pressure on central axis of the jet (at a given distance of 5.5 cm) (p; MPa)
19.4	40	2.06	10	5
19.4	60	3.09	8	6.9
19.4	95	4.90	7	10
14.3	45	3.15	8	8
14.3	70	4.90	7.5	8.3

Source: Modified from Shibazaki, M. et al., Development of oversized jet grouting, In L. F. Johansen, D. A., Bruce and M. J. Byle, eds., *Proceedings of the 3rd International Conference on Grouting and Ground Treatment*, Vol. 1, ASCE Geotechnical Special Publication 120: pp. 294–302, 2003.

from 5 to 10 MPa. Furthermore, the table shows that, for the highest rectification coefficient (4.90), the jet is more effective for the larger pipe. This is consistent with the very well-known feature of pressurised flows, for which head losses are inversely proportional to the ratio between the cross-section and the perimeter of the pipe.

Another aspect that could potentially affect the evolution of the submerged jet is represented by the hydrostatic pressure in the almost-still surrounding fluid. The first experimental researches reported in the literature on this effect are provided by Yahiro et al. (1974). Figure 3.7, taken from

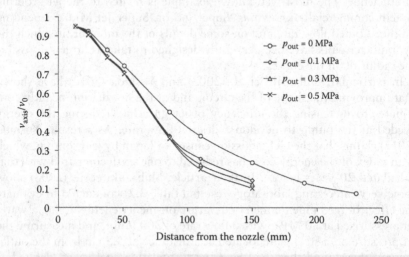

Figure 3.7 Influence of the surrounding hydrostatic pressure p_{out} on the jet velocity decay. (Adapted from Yahiro, T.H. et al. 1974. Soil improvement utilizing a high-speed waterjet and air jet. *Proceedings of the 6th Internatinal Symposium on Water Jet Technology*, Cambridge, United Kingdom, Paper J62: pp. 397–428, August 30–31, 1974.)

their work, shows the velocity decay profiles of submerged jets flowing into still water at pressures p_{out} between 0 and 0.5 MPa. Considering that the latter pressure is equivalent to a waterhead of 50 m, most of the jet grouting applications are covered by these experiments. The plot shows that the pressure of the surrounding fluid does not significantly affect the jet velocity only in the initial part of the jet, where the velocity is rather high, becoming appreciable after a distance of approximately 50 mm (which is approximately 10 times the diameter of the nozzle). The effect is significant when the pressure changes from 0 to 0.1 MPa, whereas further increases of the pressure of the surrounding fluid have a minor effect.

If the fluid jet is shrouded by a coaxial air jet, the energy losses at the jet–air interface are reduced, and as a consequence, the jet becomes more concentrated, with a smaller reduction in velocity. This effect, which forms the basic principle of double and triple fluid jet grouting systems, can be clearly seen by looking at the experiences reported by Shibazaki (1996), which are summarised in Figure 3.8. The plot shows that the velocities of the fluid along the jet axis gain a significant increment for increasing the velocities of the surrounding air jet (v_a). In particular, Shibazaki (2003) recommends adopting air velocities larger than half the sound speed (i.e., $v_a \sim 180$ m/s) to obtain the most considerable gain of cutting efficiency. However, ongoing research indicates that $v_a = 180$ m/s is a sufficient air speed.

From a qualitative point of view, the experimental evidence can be explained by considering that the shrouding air flows in the same direction of the water jet and that the air layer formed between the jet and the surrounding fluid prevents the exchange of energy between these two fluid masses and limits the effects of viscosity and turbulence. However, at some distance from the nozzle, the air jet loses its continuity and separates into bubbles that are dispersed in the fluid mass. The increase of air velocity reduces the consumption of the air layer, extending the protective effect at larger distances.

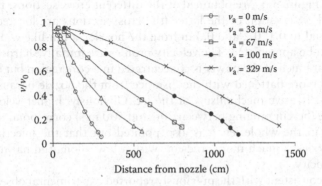

Figure 3.8 Decay of the velocity on the jet axis obtained for air-shrouded jets. (Adapted from Shibazaki, M., *Grouting and Deep Mixing* 2: 851–867, 1996.)

3.2.2 Numerical modelling

A possibility to simulate the distributions of velocities produced by different boundary conditions is given by the implementation of numerical models (e.g., Przyklenk and Schlatter 1986), where the response of complex hydraulic systems can be obtained by integrating assigned turbulence models.

An example with specific reference to jet grouting is provided by Wanik and Modoni (2012), who simulated the stationary diffusion of a submerged jet by a two-dimensional axisymmetric model. In this work, the rheological response of the injected fluid is simulated with the $k - \varepsilon$ model (Harlow and Nakayama 1968), particularly suitable for highly turbulent phenomena. The turbulent viscosity μ_t is defined by the following equation:

$$\mu_t = \rho C_\mu \frac{\kappa^2}{\varepsilon} \tag{3.2}$$

where ρ is the density of the jet fluid (kg/m^3), κ is the coefficient depending the turbulent kinetic energy (m^2/s^2), ε is its dissipation rate (m^2/s^3) and C_μ is a dimensionless constant. The calibration has been accomplished for water jets by assuming the typical values of all parameters defined for water and adjusting the dimensionless parameter C_μ (= 0.004) to reproduce the average trend observed by de Vleeschauwer and Maertens (2000) (Figure 3.4).

Some results for values of v_0 (200–500 m/s) and d (2 and 4 mm) typically adopted in jet grouting are summarised in Figure 3.9. The plot of Figure 3.9a shows the typical longitudinal profiles similar with those experimentally observed (see, e.g., Figure 3.3). The velocity, initially equal to v_0 at $x = 0$, reduces with the distance from the nozzle. Simultaneously (Figure 3.9b and c) a series of bell-shaped diagrams, similar to the experimental ones shown in Figure 3.5, are obtained at the different cross-sections. From the sequential observation of the latter (the cross-sections are located at distances equal to 0.1, 0.2 and 0.3 m from the nozzle), a fan-like widening of the jet can be appreciated. The velocity profiles are initially sharper, which means that much of the power is concentrated in a core region but becomes more and more flattened with the distance from the nozzle because of the turbulent diffusive mechanisms of the jet. Obviously, higher velocities at the nozzle (herein ranging between 200 and 500 m/s) correspond to higher velocities in the whole jet. It is also pointed out that the injection from larger nozzles is much more efficient, with a lower longitudinal reduction of the velocity.

This is consistent with the previously reported experimental observations and with their mechanical interpretation. In fact, dissipative shear stresses occur on the jet boundary, that is, on the perimeter of the jet cross-section,

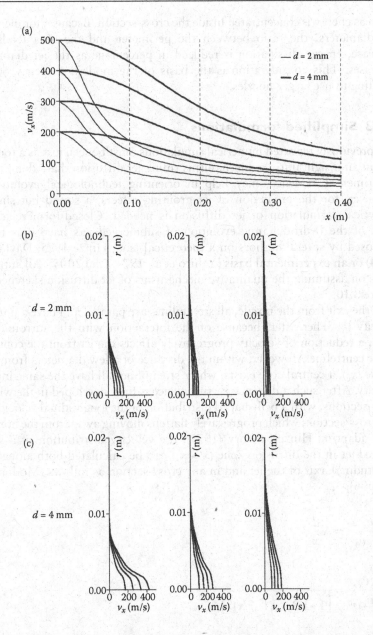

Figure 3.9 Velocity profiles calculated along the longitudinal axis of propagation (a) and at different cross-sections at $x = 0.1$ m; $x = 0.2$ m; $x = 0.3$ m (b and c) obtained for water injection with variable inlet velocities and nozzle diameters. (From Wanik, L. and G. Modoni, Numerical analysis of the diffusion of submerged jets for jet grouting application. *Incontro Annuale dei Ricercatori di Geotecnica (IARG)*, Paper 3.B.8, 6 p., 2012.)

whereas energy is concentrated inside the cross-section. Because, for increasing diameters, the ratio between the perimeter and the area tends to decrease, energy dissipation is reduced in percentage as the jet diameter increases. This consideration is the basis of the modern tendency of jet grouting to use larger nozzles.

3.2.3 Simplified formulations

The previously reported numerical simulations must be seen just as a tool to obtain an insight into the mechanisms ruling jet diffusion that, along with experimental evidence, may help in orienting technological evolutions. However, for the prediction of jet grouting effects, a sound but simple analytical formulation of jet diffusion is needed. Closed-form relationships of the hydrodynamic evolution of submerged jets have thus been proposed by several authors on a theoretical (e.g., Hinze 1975; Davidson 2004) or an experimental basis (Yahiro et al. 1974; Chu 2005). All authors agree on assuming the qualitative mechanisms of jet diffusion sketched in Figure 3.10.

At the exit from the nozzle, all streamlines are parallel and have a unique velocity v_0. Thereafter, because of the interaction with the surrounding mass, a reduction of velocity progressively affects the jet from the contour to the centreline. However, within the distance of a few diameters from the nozzle (x_c), a central core exists, where streamlines all have the same initial velocity. After such a distance x_c, turbulence is fully developed in the whole cross-sections, with a bell-shaped distribution of the longitudinal velocity in any cross-section, which progressively flattens moving away from the nozzle.

By adapting Hinze's theory (1975), the velocity distribution of a submerged jet in the diffusion zone $(x > x_c)$ can be calculated both along the longitudinal axis of the jet and in any cross-section as follows (Modoni et al. 2006):

$$\frac{v_{x,r=0}}{v_0} = \Lambda \frac{d}{x} \tag{3.3a}$$

$$\frac{v_{x,r}}{v_{x,r=0}} = \frac{1}{\left[1+1.33\cdot\Lambda^2\cdot(r/x)^2\right]^2} \tag{3.3b}$$

where r is the distance from the jet axis in the generic cross-section. By combining these two equations, the velocity in each point of coordinates (x,r) can be computed as a function of the outlet velocity v_0, of the nozzle diameter d and of a dimensionless parameter Λ that quantifies the interaction between the jet and the surrounding fluid.

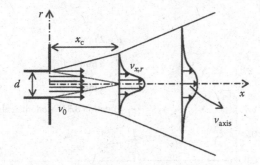

Figure 3.10 Schematic pattern of velocities in the region outside the nozzle.

The parameter Λ depends on the rheological properties of the injected and surrounding fluids, on the diameter of the nozzle and on the outlet velocity. Because differences have been observed on the decay profiles of a fluid injected with extremely different velocities (from a few metres per second to hundreds of metres per second; see Figure 3.4), Λ should be calibrated with reference to the cases of interest for jet grouting, that is, only for very high velocities (hundreds of metres per second) and very high Reynolds numbers (usually in the range $5 \times 10^4 \div 5 \times 10^6$).

The comparison reported in Figure 3.11a and b among the velocity distributions obtained with the parametric numerical analysis and those calculated by Equation 3.3 shows that, for a submerged jet of water, a fairly good fitting is obtained with $\Lambda = \Lambda_w = 16$ for the smaller nozzles ($d = 2$ and 3 mm), with a slight underestimation of the velocity distribution in the longitudinal profile obtained for the larger nozzle's diameter (4 mm). For other materials (e.g., grouts of different compositions), Modoni et al. (2006) propose to calculate Λ as follows:

$$\Lambda_g = \Lambda_w \cdot \sqrt{\frac{\mu_w}{\mu_g} \cdot \frac{\rho_g}{\rho_w}} \qquad (3.4)$$

where μ is the dynamic viscosity coefficient, and the subscripts g and w, respectively, indicate grout or water (for water, $\mu_w = 0.001$ N s/m^2 and $\rho_w = 9.81$ kN/m^3). The density of a grout ρ_g can be calculated as a function of the cement–water ratio by weight Ω, with the following equation derived from the mass balance:

$$\rho_g = \frac{\rho_c \cdot (1 + \Omega)}{\Omega + \dfrac{\rho_c}{\rho_w}} \qquad (3.5)$$

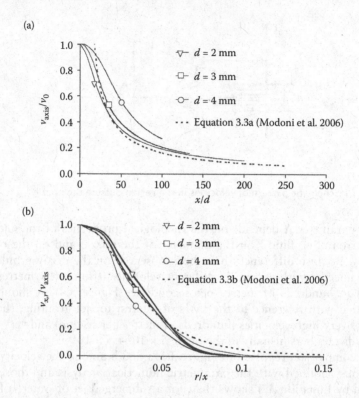

Figure 3.11 Dimensionless velocity profiles computed along the longitudinal axis (a) and in the cross-sections (b) obtained for water injection, with variable inlet velocities and nozzle diameters. (From Wanik, L. and G. Modoni, Numerical analysis of the diffusion of submerged jets for jet grouting application. *Incontro Annuale dei Ricercatori di Geotecnica (IARG)*, Paper 3.B.8, 6 p., 2012.)

where ρ_c and ρ_w represent the density of cement and water, respectively. The viscosity μ_g of a cement grout has been studied, for instance, by Raffle and Greenwood (1961; reported by Bell 1993) as a function of the same ratio Ω. Figure 3.12 summarises the values of μ_g and Λ_g for Ω ranging between 0 (i.e., water without cement) and 2 (two parts by weight of cement on one part of water).

As far as the interaction between injected and surrounding fluids is concerned, in the case of two coaxial jets (water or grout shrouded by air), the problem becomes analytically more complex: boundary conditions involve mechanisms developing both at the water (or grout)–air and air–surrounding fluid interfaces. To simplify the analysis, Flora et al. (2013) propose to consider the existence of the shrouding air jet in a simplified

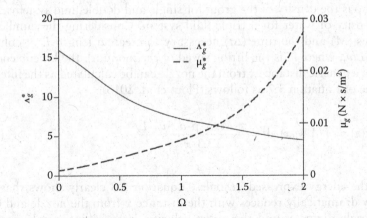

Figure 3.12 Dependency of Λ^* on grout composition (ρ_c = 3000 kg/m³; data for μ_g are taken from Raffle and Greenwood 1961, as reported by Bell 1993).

manner by modifying the value of the hydrodynamic coefficient Λ in Equation 3.3. They propose to calculate Λ as the product of two factors: one (Λ^*) related to the composition of the central jet of eroding fluid (Λ_w^* for water and Λ_g^* for grout) and the other (α_E) representing the possible influence of the shrouding air jet on boundary dissipation. In particular, $\alpha_E = 1$ for single fluid jet grouting, in which no air wrapping is given, and $\alpha_E > 1$ for double and triple fluid jet grouting. Therefore, Λ can then be written as

$$\Lambda = \alpha_E \, \Lambda^* \tag{3.6}$$

In principle, the parameter α_E introduced in Equation 3.6 should depend on the characteristics of the air jet shrouding the cutting jet (see Figure 3.8), and thus, it should be related to the air flow rate (i.e., dimensions of the nozzle and initial velocity and cross-section area of the injected air).

3.2.4 Specific energy

Given the velocity distributions in the transverse sections and along the longitudinal axis of the jet, the hydrodynamic power W of the jet can be calculated at any distance x from the nozzle via the following relation:

$$W(x) = \int_0^\infty \rho \cdot (v_{x,r})^3 \cdot \pi \cdot r \cdot dr = \frac{\pi \cdot \Lambda \cdot \rho \cdot d^3 \cdot v_0^3}{13.3 \cdot x} \tag{3.7}$$

where ρ is the density of the grout for single and double fluid systems, and the density of water for a triple fluid system. Considering the number of nozzles (M) and the time (Δt) necessary to create a length L of column ($\Delta t = L/v_r$, where v_r is the lifting speed of the monitor), the kinetic energy $E(x)$ at a generic distance x from the nozzle can be calculated as the integral in time of Equation 3.7 as follows (Flora et al. 2013):

$$E(x) = M \cdot \int_{\Delta t} W(x) \cdot dt = \frac{\pi \cdot M \cdot \Lambda \cdot \rho \cdot d^3 \cdot v_0^3 \cdot L}{13.3 \cdot x \cdot v_r} \tag{3.8}$$

with the energy expressed in joules, Equation 3.8 clearly shows that the energy dramatically reduces with the distance x from the nozzle and that the nozzle diameter and the outlet velocity, both having a cubic power exponent, play a predominant role.

A specific energy at the nozzles E_n' (joules per meter) can be defined as the kinetic energy at the nozzles per unit length of column via the following relationship:

$$E_n' = \frac{1}{2} \frac{m \cdot v_0^2}{L} = \frac{\pi}{8} \cdot \frac{M \cdot \rho \cdot d^2 \cdot v_0^3}{v_r} \tag{3.9}$$

where m represents the fluid mass ejected per unit length of column

$$\left(m = \frac{\pi}{4} \cdot \frac{M \cdot d^2 \rho \cdot v_0}{v_r} \right)$$

By combining Equations 3.8 and 3.9, the specific energy per unit length of column available at a distance x from the nozzle ($E'(x) = E(x)/L$) can be conveniently expressed as a function of the specific energy at the nozzle as

$$E'(x) = 0.6 \cdot \Lambda \frac{d}{x} \cdot E_n' \quad \text{(joules/m)} \tag{3.10}$$

which is physically consistent only when $E'(x) \leq E_n'$, that is, for $x \geq (0.6 \cdot \Lambda \cdot d)$. The specific energy defined in Equation 3.10 explicitly considers the hydrodynamic interaction between the jet and the surrounding fluid via the dimensionless parameter Λ. For such a reason, Equation 3.10 can be adapted to all jet grouting systems – single, double or triple fluid – just by calibrating Λ to consider the differences in the hydrodynamic effectiveness given by the different injection techniques. Therefore, it is a very powerful and simple tool to interpret jet grouting results.

A possible limitation in using Equation 3.10 is that all terms giving the specific energy at the nozzle (Equation 3.9) must be known, which is not always the case in practice. In fact, very often, the only known energy parameter is the so-called specific energy at the pump (Tornaghi 1989), expressed as

$$E_p' = \frac{p \cdot Q}{v_r} \tag{3.11}$$

in which p is the injection pressure at the pump, Q is the flow rate and v_r is the average monitor lifting speed. The energy calculated by Equations 3.9 and 3.11 differ because of the concentrated and distributed losses occurring in the injection pipelines (being always $E_p' \geq E_n'$). The difference between E_p' and E_n' depends on the distance between the pump and the nozzles, on the technological details of all the connections along the feed pipe, at the mast and at the jetting monitor, on the feeding pipe characteristics, as well as on the value of the applied specific energy itself E_p'. For well-designed facilities, conventional jet grouting technology and, for the typical range of values of E_p', the energy losses are usually approximately 10% of E_p' (Tornaghi 1993; Flora and Lirer 2011; AGI 2012). Hence, in the following, it can be assumed that

$$E_n' = 0.9 \cdot E_p' \tag{3.12}$$

Therefore, when only E_p' is known, Equation 3.10 can still be applied calculating E_n' via Equation 3.12.

When jet grouting is used to obtain very large columns, energies larger or much larger than usual may be adopted, and this is more and more frequent in practice; as a consequence, the energy losses may be larger than those assumed in Equation 3.12. The technological evolutions in monitor and nozzle design, however, are leading to a reduction of head losses even for extremely high energies. Therefore, if no specific information is available, Equation 3.12 may be adopted even outside the typical range of values on which it was calibrated.

3.3 JET–SOIL INTERACTION

Nowadays, several results of laboratory investigations are available on the effects of high-energy jets in different soils. In these studies, a basic distinction is made between gravelly, sandy and clayey soils, considering that different mechanisms occur depending on soil particles dimensions.

Bergschneider (2002) and Bergschneider and Walz (2003), for instance, developed a laboratory setup to visually inspect the erosion mechanism caused by high-pressure jets of water. Figure 3.13 reports a sequence of photographs taken during one of these tests on an unsaturated sandy material: soil particles are removed by the dragging action of the fluid, and a cavity is formed, progressively widening and extending in the jet direction. Simultaneously, because of the high permeability of the soil, a seepage process occurs around the cavity creating a wet zone (see the darker area in the photographs). After some time, no more expansion of the cavity is observed, which means that the erosive capacity of the jet has reached a balance with the resistance offered by the soil, but further seepage of fluid occurs in the surrounding space.

The erosive action of the jet in cohesionless soils can therefore be attributed to two concurring factors: the action of the high-speed fluid threads, which tend to drag the soil particles away from their original positions, and

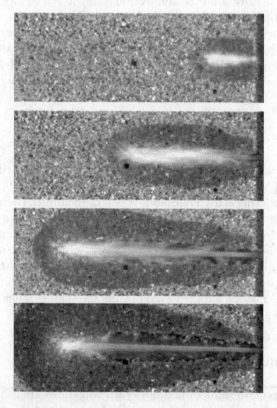

Figure 3.13 Sequence of frames taken during injection in a sandy material. (From Bergschneider B., *Zur Reichtweite beim Düsenstrahlverfahren im Sand*: PhD thesis, Bergische Universität Wuppertal Fachbereich Bauingenieurwesen [in German], 174 p., 2002.)

the pore-pressure increase of the fluid seeping around the cavity contour, which causes a reduction of the grain-to-grain contact forces.

The effect of the soil's relative density has been experimentally analysed by Stein and Graße (2003) on sandy materials having different grain size distributions. The interesting experimental evidence they noticed is that relative density affects the jet penetration rate, with a faster expansion of the cavity in looser soils (Figure 3.14). On the contrary, it was found that the limit cutting distance does not depend on the initial density. However, the cavity propagation velocity progressively reduces with jetting time, and long (in practice, unrealistic) times are thus required to reach the limit distance in the case of very dense soils. As a consequence, for the usual jetting times, relative density does play a role in the dimensions of the column. This is the first evidence, confirmed by field observations (e.g., Xanthankos et al. 1994), indicating that, it is more convenient to increase the specific jet energy than the treatment time t_j in order to have larger diameters.

A different mechanism may occur in coarse, highly permeable soils having extremely high resistance to erosion, in which permeation then becomes the leading mechanism. An example is shown in Figure 3.15, where the effect of jet grouting on a dense coarse gravelly material is reported (Modoni et al. 2008a). Figure 3.15a shows that, in the abutment of a bridge reinforced by a massive jet grouting treatment, the original layering of the soil is still clearly visible, which means that soil particles have not been removed from their position. The picture shown in Figure 3.15b confirms that, for clean gravels, the jet may not be able to completely fill the voids among the grains.

Some interesting experiments on the cutting ability of jet grouting in cohesionless soils have been reported by Yoshida et al. (1996) and Shibazaki et al. (2003). Such experiments refer to laboratory tests carried out in large tanks filled with a sandy soil, equipped to have detailed information on jet

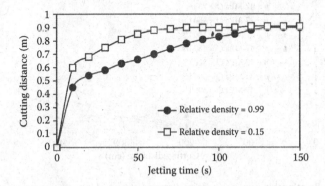

Figure 3.14 Penetration of the jetting cavity in sandy materials compacted at different relative densities. (Adapted from Stein, J. and J. Graße, Jet grouting tests and simulation, In Vanicek et al., ed., *Proceedings of the 13th ECMSGE*, Prague, Czech Republic: pp. 899–902, August 25–28, 2003.)

(a)   (b)

Figure 3.15 Effects of jet grouting on gravelly soils.

distance from the outlet nozzle. By combining different jetting times, flow rates and injection pressures, the authors have obtained the results plotted in Figures 3.16 and 3.17.

The first plot (Figure 3.16) shows that the cutting (jetting) time needed to reach a given distance, that is, to obtain a given diameter, depends on the combination of pressure and flow, and the beneficial effect of having larger flow rates for a given pressure (i.e., larger nozzle diameters) is evident. This experimental evidence is consistent with the theoretical considerations

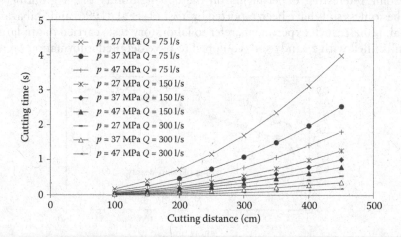

Figure 3.16 Relationship between cutting distance, time and injection parameters p and Q. (Modified from Yoshida, H. et al., Development and practical applications of large diameter soil improvement method, In Yonekura, R. and Shibazaki, M., eds., *Proceedings of the Conference on Grouting and Deep Mixing*: Tokyo, Japan: Balkema: pp. 721–726, May 14–17, 1996.)

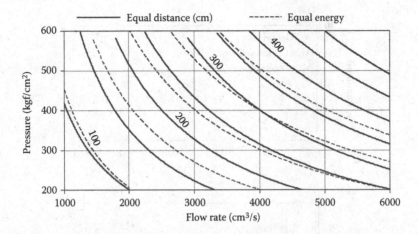

Figure 3.17 Relationship between flow rate, injection pressure and cutting distance for a given cutting time. (Modified from Yoshida, H. et al., Development and practical applications of large diameter soil improvement method, In Yonekura, R. and Shibazaki, M., eds., *Proceedings of the Conference on Grouting and Deep Mixing*: Tokyo, Japan: Balkema: pp. 721–726, May 14–17, 1996.)

previously reported in this chapter, leading to the most recent tendency to increase the diameter of the nozzle for improving jet effectiveness.

From the second plot (Figure 3.17), it is noted that the cutting distance experimentally observed can be associated with the jet energy, the curves representing these two quantities in the flow rate–pressure plot being rather similar. As a consequence, in principle, a given distance (i.e., radius of a jet column) can be obtained with different combinations of pressures and flow rates. As previously said, however, it is convenient to use the largest possible flow rates reducing the pressure, with the further beneficial effect of increasing operational safety.

In clayey soils, the effect of high-speed jets is different from cohesionless soils because adhesive interparticle forces make soil disaggregation and remoulding much more difficult. As a consequence, jet cutting produces relatively big clods of clayey material, which need to be broken by further passages of the jet if homogeneity of the jet-grouted material is sought.

Furthermore, the low permeability of these soils does not allow the injected fluid to penetrate into soil pores, and as a consequence, remoulding is the only column formation mechanism.

As an example of experimental results on fine-grained soils, Figure 3.18 (Dabbagh et al. 2002) shows that the cavity expansion velocity measured during laboratory experiments depends on the difference between the mean jet velocity and a limit velocity (v_L) (Figure 3.18a). The latter represents the minimum velocity of the fluid needed to produce cutting. It depends on the undrained shear strength of the soil (Figure 3.18b).

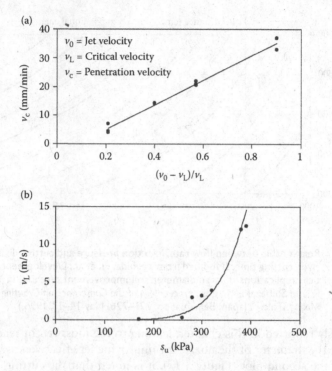

Figure 3.18 (a) Dependence of the penetration rate v_c on a limit velocity v_L; (b) limit velocity versus undrained shear strength of clayey soils. (From Dabbagh, A. A. et al., *Soils and Foundations* 42(5): pp. 1–13, 2002.)

Summarising all the evidence, the formation of a jet-grouted column can be attributed to three different jet–soil interaction mechanisms (seepage, erosion and cutting, Figure 3.19), essentially depending on soil grading.

- In gravels and sandy gravels, erosion is usually the predominant mechanism (Figure 3.19b), and the jet-grouted material is quite homogeneous. To obtain larger diameters, it is more convenient to increase the specific energy (possibly increasing the nozzle diameter and the flow rate) than the jetting time. In clean gravel, grout seepage may contribute to the formation of the column (Figure 3.19a). In this very peculiar case, it is useless to increase jet pressure or flow rate to have larger diameters, being much more effective to increase the jetting time.
- In sands, gravelly sands and silty sands, the particles are small enough to be individually removed by the dragging action of the jet, and thus erosion is the only relevant mechanism. The jet-grouted material is very homogenous because of the diffused grout and soil mixing. Since

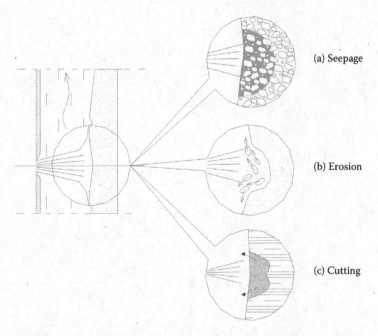

(a) Seepage

(b) Erosion

(c) Cutting

Figure 3.19 Interaction jet–soil mechanisms. (a) Seepage; (b) erosion; (c) cutting.

erosion is the only relevant formation mechanism, larger diameters can be better obtained by increasing the jet energy, specifically enlarging the nozzle diameter and the flow rate.

- In silt, sandy silt, and clay, jet cutting of soil clods is the relevant mechanism and jet-grouted material homogeneity largely depends on the number of subsequent passes at the same depth. Increasing the jetting time may then be convenient for the sake of homogeneity.

Chapter 4

Column properties

4.1 INTRODUCTION

The previous chapter has been devoted to the comprehension of the mechanisms governing jet diffusion and its interaction with different soils. In fact, a sound knowledge of these mechanical phenomena can be useful for a more rational application of the technique, allowing the selection of the most appropriate treatment procedure to create jet-grouted elements with dimensions and properties able to meet the design requirements. In practice, to this aim, it is necessary to correlate the jet grouting effects (i.e., column diameter and properties) to the original soil properties (i.e., grain size, shear strength) and to the treatment procedures (i.e., treatment parameters). However, because all soils are inherently heterogeneous, the mechanical and geometrical characteristics of the columns are usually variable. The variability of all properties can be conveniently considered using probabilistic distributions. To this aim, for a given discrete random variable x, the statistical parameters reported in Table 4.1 will be used.

4.2 TREATMENT EFFICIENCY

It is of some interest to introduce a parameter that quantifies the efficiency of treatment in terms of technical and economic convenience. In jet grouting-production, this can be done by introducing an efficiency parameter defined as the ratio between the obtained benefit (i.e., the volume per unit length of the column V_C, expressed in cubic metres per metre [m³/m], coincident with the cross-section of the column) and a parameter representing the unit cost of work, which could be the specific kinetic energy of the treatment (E') defined in the previous chapter (Equations 3.9 through 3.12). As an alternative to E', the volume of injected grout per unit length of column (V_g) expressed in cubic metres per metre (m³/m) can be considered. As a consequence, the following two efficiency parameters may be alternatively introduced (Flora and Lirer 2011; Croce et al. 2012):

Table 4.1 Statistical parameters adopted to describe randomness

Definition	Mean value (physical dimensions of x)	Variance (physical dimensions of x²)	Standard deviation (physical dimensions of x)	Coefficient of variation (dimensionless)
Formula	$\bar{x} = \dfrac{1}{n}\sum_{i=1}^{n} x_i$	$Var(X) = \dfrac{1}{n-1}\sum_{i=1}^{n}\left(x_i - \dfrac{1}{x}\right)^2$	$SD = \sqrt{Var(x)}$	$CV(X) = \dfrac{SD(x)}{x}$

$$\lambda_E = \frac{V_C}{E'} \quad \text{energetic efficiency (m}^3\text{/MJ)} \tag{4.1}$$

$$\lambda_V = \frac{V_C}{V_g} \quad \text{volumetric efficiency} \tag{4.2}$$

The reciprocal $(1/\lambda_E)$ of the energetic efficiency, called the 'volumetric specific energy' (E_s') (Tornaghi and Pettinaroli 2004), is also frequently considered in practice.

It is simple to demonstrate that, by considering the specific energy at the pump (Equation 3.11), the two parameters defined in Equations 4.1 and 4.2 are linked by the following relation:

$$\lambda_E = \frac{1}{p}\lambda_V \tag{4.3}$$

If the energy at the nozzle is considered (Equation 3.9), the following relation is obtained:

$$\lambda_E = \frac{1}{\gamma_g \dfrac{v_o^2}{2g}}\lambda_V \tag{4.4}$$

in which γ_g is the unit weight of the grout.

Larger values of the efficiency imply that a given diameter of the jet grouting column can be obtained with a lower energy input or with a lower amount of grout and, therefore, at a lower cost.

The values of the efficiency parameters λ_E and λ_V depend on both the adopted injection system and the soil grading and state. As an example, Figure 4.1 reports the energetic efficiency λ_E versus the average column diameter D_a for some field trials of single-fluid jet grouting.

The figure shows that, for a given system (in this case, single-fluid jet grouting), a clear correlation can be seen between soil gradation and jet efficiency. The coarse-grained soil is generally more prone to be eroded than the finer one, and thus, larger diameters may be consequently obtained. On the contrary, finer soils require larger treatment energies or grout volumes, and as a consequence, jet grouting in these soils is less effective. Figure 4.1 shows that the average value of λ_E in coarse soils without fines is approximately three times its value in fine-grained soils.

The larger efficiency of jet grouting in sands and sandy gravels also implies that the amount of grout spoil is the lowest for them, as experimentally confirmed by several authors (e.g., Kauschinger et al. 1992; Covil and Skinner 1994). A second interesting consideration is that, although the energies to be used on site may vary in a wide range for a given jet grouting technology, depending on the desired column diameter, the efficiencies vary in a much narrower range, which essentially depends on the soil properties. This is the reason why efficiencies may be conveniently adopted as a simple means for the preliminary estimate of the cost of jet grouting. Tables 4.2 and 4.3 summarise, typical values of the energetic efficiency λ_E for the cases of single- and double-fluid jet grouting, respectively. In practice, values external to the ranges reported in the tables may be possible because

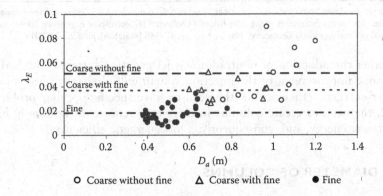

Figure 4.1 Energetic efficiency λ_E calculated from some field trials of single-fluid jet grouting. The horizontal lines represent the average values for the three considered classes of soils. (Modified from Flora, A. et al., The diameter of single-, double- and triple-fluid jet grouting columns: Prediction method and field trial results. *Géotechnique* 63(11): pp. 934–945, 2013.)

Table 4.2 Typical values of the energetic efficiency λ_E and of its reciprocal $1/\lambda_E$ (volumetric specific energy) for a single-fluid jet grouting treatment in different soils

Soil	Energetic efficiency λ_E (m³/MJ)	Volumetric specific energy $1/\lambda_E$ (MJ/m³)
Sandy gravel	0.067–0.100	10–15
From gravelly sand to silty sand	0.033–0.067	15–30
From sandy silt to clayey silt (low consistency)	0.020–0.033	30–50
From sandy silt to clayey silt (high consistency)	<0.020	>50

Source: Modified from Flora, A. and S. Lirer, Interventi di consolidamento dei terreni, tecnologie e scelte di progetto. *Proceedings of the 24th National Conference of Geotechnical Engineering 'Innovazione tecnologica nell'Ingegneria Geotecnica'*, Napoli, Italy: pp. 87–148 [in Italian], June 22–24, 2011.

Table 4.3 Typical values of the energetic efficiency λ_E and of its reciprocal $1/\lambda_E$ (volumetric specific energy) for a double-fluid jet grouting treatment in different soils

Soil	Energetic efficiency λ_E (m³/MJ)	Volumetric specific energy $1/\lambda_E$ (MJ/m³)
From sandy gravel to silty sand	0.077–0.125	8–13
From sandy silt to clayey silt (low consistency)	0.077–0.025	13–40
From sandy silt to clayey silt (high consistency)	<0.025	>40

Source: Modified from Flora, A. and S. Lirer, Interventi di consolidamento dei terreni, tecnologie e scelte di progetto. *Proceedings of the 24th National Conference of Geotechnical Engineering 'Innovazione tecnologica nell'Ingegneria Geotecnica'*, Napoli, Italy: pp. 87–148 [in Italian], June 22–24, 2011.

of either the adoption of nontraditional jet grouting systems (any hydrodynamic improvement of the injection system will obviously result into a larger energetic efficiency) or the need to solve specific peculiar problems (e.g., the need to create columns in clays may require an unusually high treatment energy, and, consequently, a low energetic efficiency).

4.3 DIAMETER OF COLUMNS

4.3.1 Experimental evidence

The first step of any jet grouting project is the prediction of the diameter of the columns. In fact, static or waterproofing functions can be effectively guaranteed only if the columns possess the desired dimensions and, as a consequence, the geometrical assembly of the columns complies with the design hypotheses.

As shown in Chapter 3, the diameter of the columns depends on jet hydrodynamic properties as well as on soil resistance to erosion. It is thus important to choose the most appropriate technique by selecting jet grouting system and treatment parameters according to the soil to be treated and to the diameter required.

Clear evidence of the influence of treatment procedure and soil properties on the diameter columns is given in Figures 4.2 and 4.3 (Croce and

Figure 4.2 Columns obtained in coarse-grained pyroclastic soils by the single-fluid system with different treatment parameters. (Modified from Croce, P. and A. Flora, *Rivista Italiana di Geotecnica* 2: pp. 5–14 [in Italian], 1998.)

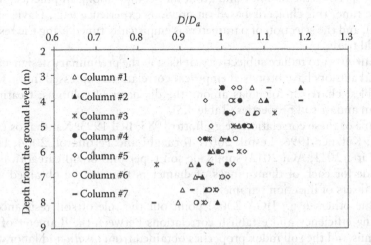

Figure 4.3 Diameter variation with depth. (Modified from Croce, P. and A. Flora, *Rivista Italiana di Geotecnica* 2: pp. 5–14 [in Italian], 1998.)

Flora 1998), which shows seven single-fluid jet-grouted columns obtained in a coarse-grained pyroclastic soil, with the different treatment parameters reported in Table 4.4. The figure has been plotted in· a nondimensional form, because the different columns have different average diameters due to the various combinations of treatment parameters. The progressive reduction with depth can be attributed to the increase in soil shear strength caused by the increase in the overburden effective stress. This effect may be relevant only in soils having high values of the friction angle (in the case reported in the figure, $\varphi' = 39°$). In soils having lower friction angles, the reduction of diameter with depth is usually negligible. Figure 4.3 also shows that there is a random variation of diameter at any depth due to the inherent variability of soil properties.

It is thus suggested to consider the diameter as a stochastic variable having some probability distribution. The mean value D_m can be expressed as a function of treatment parameters, soil properties and stress state. The random deviation from the mean is considered as the result of a stochastic process. Consistently with this logical path, indications in this chapter are first given to predict the mean value of the diameter of columns; then, a statistical approach is proposed to quantify the observed scatter.

4.3.2 Mean diameter

The desired value of the mean diameter of columns can be obtained by selecting, for the specific soil to be treated, an appropriate injection technique (single, double or triple fluid) and a proper set of treatment parameters (number and diameter of nozzles, injection pressure and/or flow rate, lifting speed of the monitor and grout mix composition). In practice, most of the time, this choice is based on previous experience (e.g., Davie et al. 2003), and the selection of parameters is empirically tuned using the results of field trials.

In an effort to reduce subjectivity, at least at the preliminary design stage, several authors have proposed empirical correlations expressed in the form of tables, charts or formulas among the diameter of columns, treatment system and/or soil properties (Table 4.5).

Some of these correlations (e.g., Botto 1985; Bell 1993; Xanthakos et al. 1994; Kutzner 1996; Lesnik 2001; Tornaghi and Pettinaroli 2004; Flora and Lirer 2011; AGI 2012) subdivide soil types into broad categories and provide for each of them ranges of diameters that can be obtained with typical sets of injection parameters. '

Some others (e.g., JJGA 2005) point out the role of soil resistance in eroding efficiency and establish correlations between the diameter of the columns and the soil index properties obtained from *in situ* and laboratory tests.

Table 4.4 Treatment parameters adopted for the columns of Figure 4.1

Column no. (from left to right in Figure 4.2)	Injection pressure (MPa)	Flow rate (10^{-3} m³/s)	Number of nozzles	Diameter of nozzles (mm)	Rotational speed (rad/s)	Average lifting speed (mm/s)	Volume of grout per unit length of column (m³/m)	D_a (m)
1	45	2.50	1	3.8	0.94	4.00	0.625	0.95
2	45	2.50	1	3.8	0.63	4.00	0.625	1.11
3	45	2.35	2	2.6	1.05	6.67	0.353	0.71
4	45	2.50	1	3.8	1.18	5.00	0.500	0.97
5	45	2.35	2	2.6	1.57	6.67	0.353	0.75
6	45	2.50	1	3.8	0.79	5.00	0.500	0.96
7	45	1.38	2	2.0	1.57	5.71	0.242	0.66

Table 4.5 List of correlations available in literature

Reference	Form of correlation	Soil classification
Botto (1985), Bell (1993)	Chart	Soil type
Miki and Nakanishi (1984), Shibazaki (1996)	Chart	Coarse grained (N_{spt})
Xanthakos et al. (1994)	Table	Fine to coarse grained
Kutzner (1996)	Table	Soil type
Tornaghi (1989)	Chart	Coarse grained (N_{spt}); fine grained (s_u)
JJGA (2005)	Table	N_{spt}
Tornaghi and Pettinaroli (2004), Flora and Lirer (2011)	Chart	Soil type
AGI (2012)	Table	Soil type
Modoni et al. (2006), Croce et al. (2011)	Chart	Shear strength parameters (φ', s_u)
Wang et al. (2012)	Equation	Soil type
Flora et al. (2013)	Equation, charts	Coarse grained (N_{spt}); fine grained (q_c)

Most of these correlations do not explicitly consider treatment parameters but simply refer to the different jet grouting systems. Some authors (e.g., Tornaghi 1989, 1993; Croce and Flora 2000; Tornaghi and Pettinaroli 2004; Flora and Lirer 2011; Flora et al. 2013) propose more comprehensive relationships, including the combined effect of soil properties and treatment parameters.

Tornaghi (1989) introduces the specific energy (i.e., the energy per unit length of column) supplied at the pump E'_p (see Equation 3.11) as the meaningful variable to summarise all treatment parameters. Croce and Flora (2000) argue that the concentrated and distributed energy losses occurring from the pump to the nozzles must be discounted and thus propose to use the kinetic energy at the nozzle per unit length of column as the reference parameter (Equation 3.9). Flora et al. (2013) introduce a more consistent formulation of the specific kinetic energy of the jet at a distance x from the nozzle (Equation 3.10), thus allowing a more reliable prediction of the mean diameter. This latter approach will be described in detail in the final part of this section.

As an alternative to empirical correlations, theoretical models able to describe the hydrodynamic mechanisms occurring during the diffusion of submerged jets and the impact on the soil have been proposed by Chu (2005), Modoni et al. (2006), Heng (2008) and Wang et al. (2012). The theoretical approach has the advantage of being based on a more accurate analysis of the phenomena, explicitly considering the role of each variable (either soil property or treatment parameter). However, because of the complexity of the phenomena induced by jet grouting, theoretical models are highly simplified and have a number of parameters to be necessarily calibrated on experimental data.

4.3.2.1 Simplified prediction of mean diameter

A simple prediction of columns diameter, useful to evaluate the feasibility of jet grouting at a preliminary design stage, can be made by considering any of the empirical correlations previously mentioned. As an example, Table 4.6 (AGI 2012) reports typical diameter ranges for different soils and treatment systems.

A simple, alternative way to estimate the mean diameter of jet-grouted columns at the early stage of design, when detailed information on soil properties is not available, is to use the previously introduced concept of treatment efficiency. In fact, once the jet grouting system has been chosen, the mean diameter of jet-grouted columns can be evaluated by associating to each class of soils a typical range of energetic efficiencies λ_E (as reported in Tables 4.2 and 4.3 for single- and double-fluid jet grouting, respectively). Then, combining Equations 4.1 to 4.4, the mean diameter of jet-grouted columns can be expressed as

$$D_{mean} = 1.128\sqrt{p \cdot V_g \cdot \lambda_E} \tag{4.5}$$

in which, for dimensional consistency, pressure p must be expressed in MPa; the volume of injected grout V_g, in m^3/m; and the energetic efficiency λ_E, in m^3/MJ.

Equation 4.5 has been used with some possible values of the treatment parameters and of the energetic efficiency to draw the chart reported in

Table 4.6 Typical values of the mean diameter of jet grouting columns

Treatment system	Mean diameter of columns (m)			
	Moderately stiff clay	Soft silt and clay	Silty sand	Sand and/or gravel
Single fluid	NR*	0.4–0.8	0.6–1.0	0.6–1.2
Double fluid	0.5–1.0	0.6–1.3	1.0–2.0	1.2–2.5
Triple fluid	0.8–1.5	1.0–1.8	1.2–2.5	1.5–3.0

Source: AGI, *Jet Grouting Guidelines*: Associazione Geotecnica Italiana [in Italian], 69 pp., 2012.

*NR, not recommended.

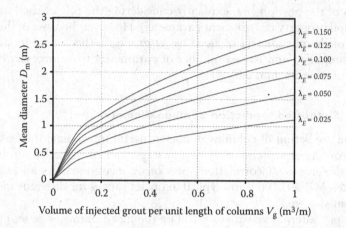

Figure 4.4 Mean diameter of the column versus the volume of injected grout per unit length V_g for different values of the energetic efficiency λ_E (see Tables 4.2 and 4.3) and for $p = 40$ MPa.

Figure 4.4. It may be worth noticing that, in Equation 4.5, the influence of soil properties and jet grouting system is hidden in the value of the energetic efficiency λ_E (see Tables 4.3 and 4.4).

4.3.2.2 Advanced prediction of mean diameter

In an attempt to mix the benefits of theory and empiricism into a unique approach, Flora et al. (2013) propose to calculate the mean diameter of a jet grouting column D_m obtained with any system (single, double or triple fluid) using a simple expression that explicitly considers jet energy and soil resistance. The analytical formulation stems from the consideration that the average diameter must be directly proportional to the jet erosive capacity, expressed via the specific jet energy $E'(x)$ (Equation 3.10), and inversely proportional to soil resistance to erosion (S). Formally, the following relation must hold:

$$D_m \propto E'(x)^\beta \cdot S^\delta \qquad (4.6)$$

with the β and δ (that must be negative) exponents accounting for the non-linearity of the relationship among D_m, $E'(x)$ and S, being x the radial distance from column axis. In the following, the two exponents β and δ will be considered constant. Equation 4.6, it can be written for a reference column diameter D_{ref} obtained using a reference specific kinetic energy $E'_{ref}(x)$ in a soil having a reference resistance to the erosive capacity S_{ref}. Then, whatever the coefficient of proportionality between D_a and the group $E'(x)^\beta \cdot S^\delta$, it can be written that

$$\frac{D_{mean}}{D_{ref}} = \left(\frac{E'(x)}{E'_{ref}(x)}\right)^\beta \cdot \left(\frac{S}{S_{ref}}\right)^\delta \qquad (4.7)$$

The two exponents β and δ should be calibrated on the widest possible set of experimental data. Considering Equation 3.10, the ratio between the jet kinetic energy and its reference value can be written as

$$\frac{E'(x)}{E'_{ref}(x)} = \frac{\alpha_E \cdot \Lambda^* \cdot E'_n}{\alpha_{Eref} \cdot \Lambda^*_{ref} \cdot E'_{n,ref}} \qquad (4.8)$$

As shown in the previous chapter, soil resistance to erosion depends on its shear strength. Therefore, it can be simply expressed considering the results of popular geotechnical *in situ* tests (namely, standard penetration test [SPT] and cone penetration test [CPT], having as results, respectively, the blow count N_{SPT} and the tip penetration resistance q_c). Assuming that, in coarse-grained soils, it is typical to perform SPTs, whereas in fine-grained soils, it is more typical to carry out CPTs, the ratio S/S_{ref} of Equation 4.7 can be written as

$$\frac{S}{S_{ref}} = \frac{N_{SPT}}{N_{SPT_{ref}}} \quad \text{for coarse-grained soils} \qquad (4.9a)$$

$$\frac{S}{S_{ref}} = \frac{q_c}{q_{c_{ref}}} \quad \text{for fine-grained soils} \qquad (4.9b)$$

The reference parameters $N_{SPT_{ref}}$ and $q_{c_{ref}}$ are those corresponding to a column having a diameter D_{ref} with a jet grouting treatment carried out with the specific energy E'_{ref}. The reference kinetic energy $E'_{ref}(x)$ has been calculated with reference to a single-fluid jet grouting treatment (thus, considering $\alpha_{E\_ref} = 1$), with a cement–water ratio by weight ω equal to 1 ($\Lambda^*_{ref} \approx 7.5$) and with a specific energy at the nozzles of $E'_{n,ref} = 10$ MJ/m (Flora et al. 2013). The proposed reference values for the soil shear strength are $N_{SPT_{ref}} = 10$ and $q_{c_{ref}} = 1.5$ MPa. As will be shown in the following, D_{ref} can be calibrated on the results of field trials.

Finally, combining all these considerations and introducing them into Equation 4.7, the mean diameter D_m can be rewritten as

$$D_m = D_{ref} \cdot \left(\frac{\alpha_E \cdot \Lambda^* \cdot E_n'}{7.5 \cdot 10} \right)^\beta \cdot \left(\frac{q_c}{1.5} \right)^\delta \quad \text{(for fine-grained soils,} \atop E_n' \text{ in MJ/m and } q_c \text{ in MPa)} \qquad (4.10a)$$

$$D_m = D_{ref} \cdot \left(\frac{\alpha_E \cdot \Lambda^* \cdot E_n'}{7.5 \cdot 10} \right)^\beta \cdot \left(\frac{N_{SPT}}{10} \right)^\delta \quad \text{(for coarse-grained soils,} \atop \text{with } E_n' \text{ in MJ/m)} \qquad (4.10b)$$

where the values of Λ^* are assigned (Figure 3.12) depending on the cement–water ratio by weight of the cutting fluid. For triple-fluid jet grouting, in which water (without cement) is used as cutting fluid ($\Omega = 0$), the value $\Lambda^* = 16$ has been assigned in Equations 4.10a and 4.10b.

The calibration of the other parameters of Equations 4.10a and 4.10b has been accomplished by collecting, from the literature or from the personal experience of the authors, data on a number of field trials of single-, double- and triple-fluid jet grouting (a detailed report of these data can be found in Flora et al. 2013).

In most cases, the velocity of the air jet, which plays a role (see Chapter 3.2.1 and Figure 3.8), is not known, and as a consequence, the effect of air shrouding has been considered in a simplified manner, considering a single, constant value of the parameter α_E. Improvements to the prediction of the mean diameter will be obtained when experimental information on the air jet properties will be known in a significant number of cases.

The parameters D_{ref}, β, δ and α_E of Equations 4.10a and 4.10b have been calibrated by best fitting all the available experimental data (Figure 4.5). Since silty sands and gravels show different sensitivities to erosion, a further subdivision has been introduced on coarse-grained soils, distinguishing those having a significant content of fine particles (with fine) from those having a negligible amount (without fine). Accordingly, two different values of D_{ref} have been found for each of these categories in the calibration of Equation 4.10b.

Then, the collected data have been grouped in three main categories: fine-grained materials, coarse-grained materials with fine and coarse-grained materials without fine. When a range of soil resistance values (either q_c or N_{SPT}) was available, the mean value was used in Equations 4.10a and 4.10b.

The set of best-fitting parameters (imposing a unique value of β and δ) is reported in Table 4.7.

As shown in Table 4.7, the values of D_{ref} adopted in the calibration are consistent with the experimental evidence, reducing as the content of fine in the soil increases.

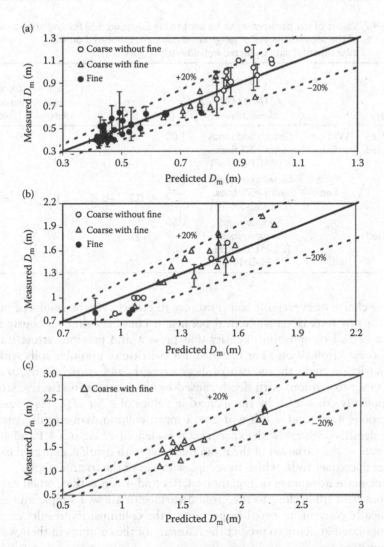

Figure 4.5 Comparison between the measured (experimental) and predicted (Equations 4.10a and 4.10b) values of the mean diameter of a jet grouting column for single-fluid (a), double-fluid (b) and triple-fluid (c) jet grouting.

The resulting comparison between measured and predicted mean diameters (Figure 4.5a, b and c) shows a satisfactory agreement, the scatter being mostly contained in a range of ±20%. It is worth noticing that, for the case of single fluid, where a statistical interpretation of the experimental data could be carried out, the scatter is similar to the dispersion of measured data (the vertical bars in Figure 4.5a represent the values of the standard deviation of the measured diameters).

Table 4.7 Values of the parameters to be adopted in Equation 4.10 for the prediction of the mean diameter of jet-grouted columns, calibrated on the experimental data collected in field trials (Figure 4.5)

Soil type		ASTM D2487 classification	D_{ref} (m)	β	δ	α_E (single fluid)	α_E (double and triple fluid)
Coarse grained	Without fine	Gravels and sands with <5% fines (GW-GP-SW-SP)	1.00				
	With fine	Gravels and sands with >5% fines (GM-GC-SM-SC)	0.80	0.2	−0.25	1	6
Fine grained		Silts, clay and organic soils (CL-ML-OL-CH-MH-OH-Pt)	0.50				

The choice of expressing soil resistance to erosion via the results of popular *in situ* tests is very practical because it considers simultaneously the effect of soil composition, density and present and previous stress states (i.e., overconsolidation). For instance, for the case of granular soils with a given friction angle, the increase of shear strength (and, consequently, of soil resistance to erosion) with depth caused by the increase in effective stress is implicitly considered by the increasing values of q_c or N_{SPT}. As a result, Equations 4.10a and 4.10b will give a mean column diameter decreasing with depth, consistent with experimental evidence (see Figures 4.1 and 4.3). The rate of the variation of the mean diameter with depth is expected to be larger in coarser soils, which have higher values of the friction angle.

One main advantage of Equations 4.10a and 4.10b is that, when carrying out field trials close to the ground surface to allow a simple and economically convenient visual inspection of the columns, the results can be extrapolated at depth to predict the diameter of the columns in their working position.

Based on the above equations, design charts can be produced (Figures 4.6, 4.7 and 4.8), reporting, for coarse- (with and without fine) and fine-grained soils and for single-, double- and triple-fluid jet grouting, the mean diameter D_m expressed as a function of the soil properties (N_{SPT} and q_c) and of the kinetic energy at the nozzle per unit length of column E'_n.

The plots for single- and double-fluid systems have been computed for a cement–water ratio $\Omega = 1$ (and, therefore, $\Lambda^* = 7.5$ in Equations 4.10a and 4.10b); for grouts of different composition, a correcting factor $(\Lambda^*/7.5)^\beta$ should be applied to the diameter of Figures 4.6 and 4.7, considering the value Λ^* reported in Figure 3.12 as a function of Ω.

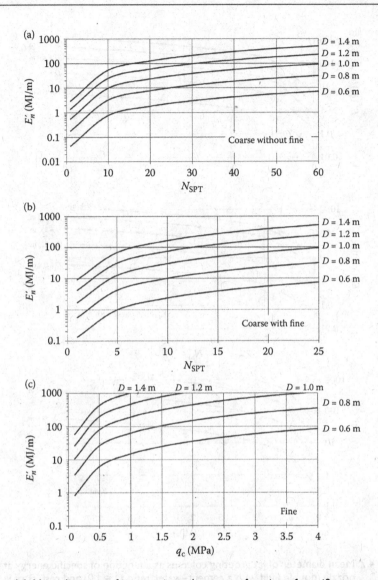

Figure 4.6 Mean diameter of jet grouting columns as a function of specific energy at the nozzle (single fluid with a cement–water ratio $\Omega = 1.0$) and results of *in situ* tests. (a) Coarse without fine, (b) coarse with fine, (c) fine.

For triple fluid, in which water is the cutting fluid, the diameters have been computed assigning $\Lambda^* = 16$ in Equations 4.10a and 4.10b.

Figures 4.6, 4.7 and 4.8 (or, equivalently, Equations 4.10a and 4.10b) can be useful at the design stage. The equations have been calibrated on field trials with energies not higher than 100 MJ/m and should be used with

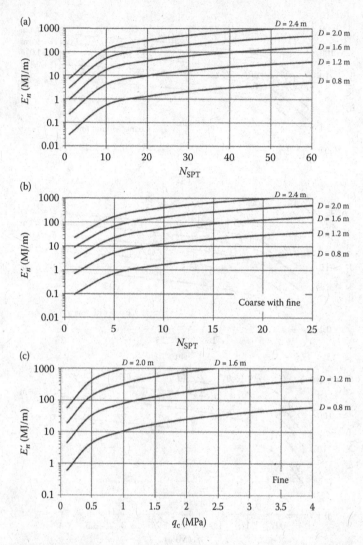

Figure 4.7 Mean diameter of jet grouting columns as a function of specific energy at the nozzle (double fluid with a cement–water ratio $\Omega = 1.0$) and results of *in situ* tests. (a) Coarse without fine, (b) coarse with fine, (c) fine.

extreme care out of the calibration range. In particular, it is expected that at higher energies, as adopted in the most recent practice due to technological evolutions, the proposed formulation underestimates the diameter for a given value of the specific energy. Research is ongoing to have a calibration best suited for such high energies. For a given soil, knowing q_c or N_{SPT}, the desired column diameter D_m can be obtained by selecting the desired jet grouting system. A corresponding value of the specific kinetic energy E'_n

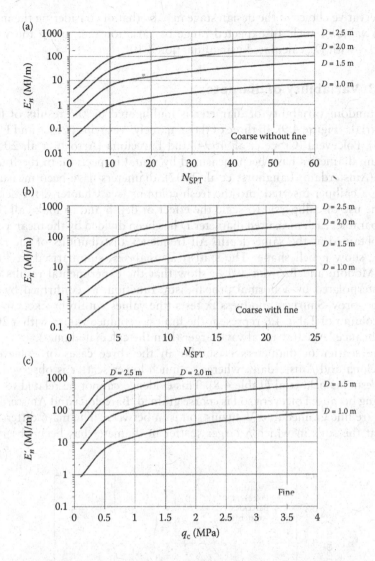

Figure 4.8 Mean diameter of jet grouting columns as a function of specific energy at the nozzle (triple fluid) and results of *in situ* tests. (a) Coarse without fine, (b) coarse with fine, (c) fine.

will then be calculated, and a set of jetting parameters have to be assigned, which combine to give the value E'_n. Improvements to the design charts for double and triple fluid systems may be obtained in the future by considering different values of α for different values of air velocity.

Since by using Equation 3.10 a difference between the calculated and measured diameter of the column of up to ±20% can be expected, a conservative choice at the design stage may be that of considering the minimum value of such an expected range of variation (i.e., using the value $0.8 \cdot D_m$, with D_m calculated using Equation 3.10).

4.3.3 Variability of diameter

The random variability of diameter is highlighted by the results of four field trials (Figure 4.9). In three of them, namely, Vesuvius (Croce and Flora 1998), Polcevera (Croce et al. 1994) and Barcelona (Arroyo et al. 2007), column diameters have been measured by visual inspection; in the fourth case (Amsterdam; Langhorst et al. 2007), diameters have been measured using a calliper inserted into the fresh columns (see Chapter 8 for details on the use of callipers). To scale the effect of depth and to make all data comparable, each measured diameter D has been divided by the mean value D_m obtained at the same depth. All frequency distributions of the ratio D/D_m show a bell shape. The statistical analyses, summarised in Table 4.8 (Modoni and Bzówka 2012), show that the experimental results can be interpolated by a normal probabilistic function, as confirmed by the Kolmogorov–Smirnov goodness fit tests (the values within brackets in the last column of Table 4.6 represent the limit acceptance values with a 20% significance level and are always larger than the calculated ones).

The scatter of diameters is similar in the three cases of Polcevera, Barcelona and Amsterdam, whereas a much lower scatter is observed for the Vesuvius field trial (Table 4.8). This evidence cannot be related to soil grading because Polcevera soil is coarse grained, Barcelona and Amsterdam soils are fine grained and Vesuvius soil is in between. The main difference is that the soils in which a larger scatter in diameter has been observed

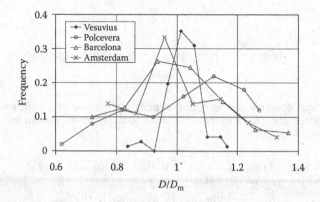

Figure 4.9 Frequency distributions of columns diameters from different field trials. (Adapted from Modoni, G. and J. Bzówka, *ASCE Journal of Geotechnical and Geoenvironmental Engineering* 138(12): pp. 1442–1454, 2012).

Table 4.8 Statistical analysis of column diameters

Case study	Soil	Number of data	$\bar{D}_{min} - \bar{D}_{max}(m)$	$CV(D/D_m)$	Kolmogorov–Smirnov test
Vesuvius	Pyroclastic silty sand	71	0.55–1.35	0.06	0.09 (0.13)
Polcevera	Alluvial sandy gravel	50	1.06–1.20	0.19	0.12 (0.15)
Barcelona	Alluvial sandy clay	97	0.35–0.64	0.18	0.06 (0.11)
Amsterdam	Stratified sandy clay and clay	72	0.72–1.37	0.16	0.10 (0.13)

Source: Adapted from Modoni, G. and J. Bzówka, ASCE Journal of Geotechnical and Geoenvironmental Engineering 138(12): pp. 1442–1454, 2012.

(Polcevera, Barcelona and Amsterdam) are more heterogeneous than the Vesuvius soil: in fact, Polcevera and Barcelona are alluvial soils and are thus intrinsically heterogeneous, and Amsterdam subsoil, in the area of the field trial, is stratified with continuous variations in grading and is, therefore, highly heterogeneous too.

Then, these field trial results indicate that soil heterogeneity is the ruling factor on the scatter of diameters. On the basis of a larger collection of experimental information, ranges of the expected values of CV(D) are reported in Table 4.9 as a function of a qualitative indication of soil heterogeneity.

Often, soil heterogeneity is connected to soil grading: for instance, thick clayey deposits are expected to be homogenous (therefore, columns will have low values of CV(D)), whereas coarse-grained soils are usually heterogeneous (higher values of CV(D)). The variability of the results of *in situ* tests (e.g., CPT tip penetration resistance q_c) may give a hint on this, as well as good knowledge of the geological origin of the soil deposit.

Based on the indications in Table 4.9, it is possible to derive an estimate of the diameter associated to an acceptable probability of having a value lower than the estimated one. As an example, by setting a probability P equal to 5% (which is typical of many applications in civil engineering), the corresponding design value ($D_{5\%}$) can be simply calculated as

$$D_{5\%} = D_m \cdot [1 - g(n) \cdot CV(D)] \tag{4.11}$$

Table 4.9 Coefficients of variation CV(D) of the diameter of columns for soils without significant discontinuities

	Soil heterogeneity		
	Low	Medium	High
CV(D)	0.02–0.05	0.05–0.10	0.05–0.20

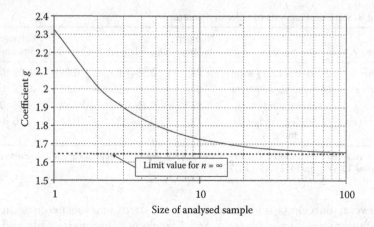

Figure 4.10 Numerical value of the coefficient g(n) corresponding to the 5% fractile in the Gauss distribution as a function of the number of available experimental data (Equation 4.12).

in which n is the number of elements (in our case, the experimental measurements) of the analysed sample of data, and $g(n)$ is a function defined as

$$g(n) = \left(1.645 \cdot \sqrt{\frac{1}{n} + 1} \right) \tag{4.12}$$

In the case of jet grouting, for which, most times, a limited number n of experimental measurements is available, it is important to stress out that the coefficient $g(n)$ assumes the well-known value $g = 1.645$ only asymptotically. Figure 4.10 shows that the coefficient $g(n)$ (and, therefore, the difference between $D_{5\%}$ and D_m) can increase significantly when n is low, and this must be considered in processing experimental results.

4.4 DEVIATION OF THE COLUMNS AXIS

A relevant problem of jet grouting is the possible deviation of the column axis from its design position. The magnitude of such a deviation depends on several factors, such as equipment quality, accuracy of the rig placement and of the treatment procedure, subsoil characteristics, perforation direction and length.

The deviation of the column axis may significantly affect the performance of those structures for which the continuity is a fundamental requirement (e.g., cutoff barriers, bottom plugs, provisional lining of tunnels). Deviation can be reduced to tolerable levels, and countermeasures can be adopted, provided that an accurate monitoring system is installed on the drilling equipment. A review of the techniques and the instruments adopted to measure and control the deviation of column axes will be shown in Chapter 8.

However, experimental measurements (e.g., Croce and Modoni 2005) indicate that a deviation of the axis of the column is unavoidable, even in carefully controlled conditions, and as a consequence, it should be considered as a physiological defect of jet grouting.

An example of measurement, performed on a cutoff made of a double barrier of partially overlapped 16-m-long columns (executed from a depth of 21 to 37 m), is reported in Figure 4.11. In this case, the inclination of each column axis was measured by means of an inclinometer located at the lower end of the perforation tool. The first plot (Figure 4.11a) shows the projection of the bottom tip of the column axes on a horizontal plane. If the axes were perfectly vertical, all the dots would coincide with the origin of

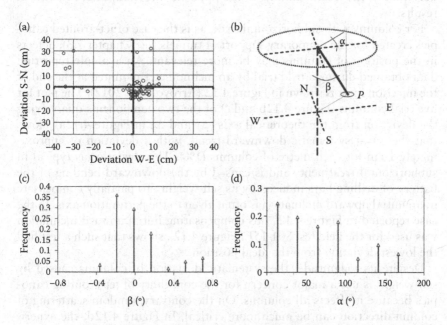

Figure 4.11 Deviation from the verticality of columns. (a) Position of the tip of the monitor at the bottom end of the column, (b) definition of the inclination β and the azimuth α, (c) and (d) values of β and α calculated from the data reported in plot a. (Modified from Croce, P. and G. Modoni, Ground Improvement 10(1): pp. 1–9, 2005; Croce, P. and G. Modoni, *Ground Improvement* 13(1): pp. 47–50, 2008.)

the diagram. The maximum observed deviation is equal to 45 cm, although most of the deviations reported in the figure are much lower.

A convenient way to define the deviation of columns from the design position is shown in Figure 4.11b, introducing the azimuth (α) and the inclination (β), with $0° \leq \alpha \leq 180°$ and $-90° \leq \beta \leq +90°$. The direction of columns is defined by α and β. Figure 4.11c and d show the frequency distributions of the two angles, with reference to the experimental case considered in the figure. The first plot shows that the inclination angle β is almost symmetrically distributed around zero, with a standard deviation of 0.27°, and that a bell-shaped curve can be approximately assumed to represent its variation. This is consistent with the experimental data previously reported by Croce et al. (2004a).

On the contrary, a rather uniform frequency distribution is observed for the azimuth angle α. This is physically consistent because, in the case of vertical columns, there is no reason why a single value of α should be more probable than others, unless systematic errors (one for all: the existence of an already cemented body on one side, which introduces a systematic deviation of the treatment axis in the opposite direction) affect treatment results.

For columns with subhorizontal axes, as is the case of jet-grouted canopies created for the temporary support of tunnels (see Chapter 7.3), defects in the position of columns may be more relevant. An example reporting data obtained during a field trial by an inclinometer mounted at the end of the injection mast is shown in Figure 4.12 (Arroyo et al. 2012). The plot for five trial columns (Figure 4.12b and c) of the two components quantifying the deviation from the theoretical axis (Δy and Δz in Figure 4.12a) shows that there is a systematic downward trend, with a deviation of approximately 1 cm for each metre of columns (1%). This deviation is typical in subhorizontal treatments and is caused by the downward bending of the battery of drilling bars induced by its self-weight. To partially compensate it, an initial upward inclination is often given to the perforation axis. In the case reported in Figure 4.12, this compensating initial upward inclination was used for the field test S01-15E. Figure 4.12a shows that such a test had the lowest deviation from the ideal position.

Despite its magnitude, the systematic deviation of columns caused by self-weight is not a major concern for the continuity of jet grouting canopies because it affects all columns. On the contrary, random scattering of column direction can be much more critical. In Figure 4.12d, the experimentally obtained frequency distribution of β shows a bell-shaped curve, almost symmetrical around the value $\beta = 0$. The standard deviation calculated in this field trial is low ($SD(\beta) = 0.17°$), whereas the one calculated in one of the canopies during tunnel construction is significantly larger ($SD(\beta) = 0.68°$).

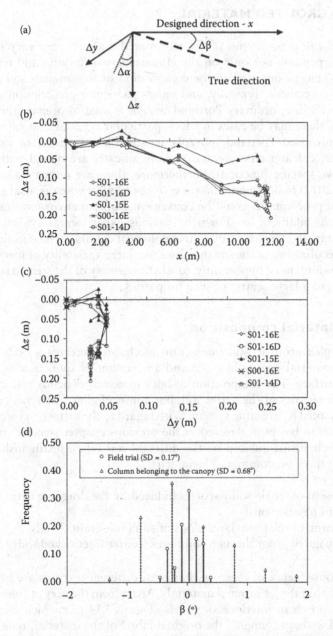

Figure 4.12 Orientation of column axes in tunnel canopies. (Modified from Arroyo, M. et al., Design of jet grouting for tunnel waterproofing, In Viggiani, ed., *Proceedings of the 7th International Symposium on the Geotechnical Aspects of Underground Construction in Soft Ground*: TC28-IS Rome: London, United Kingdom: Taylor & Francis Group: pp. 181–188, 2012.)

4.5 JET-GROUTED MATERIAL

The mechanical properties of the jet-grouted material may vary in wide ranges, depending not only on the effectiveness of mixing and replacing soil with cement but also on the composition of the original soil and on the adopted cement. Typically, and unless particular prescription is given for a specific site, ordinary Portland cement is used to prepare the grout. However, there may be cases in which particular types of cement are used (finely pulverised Portland, pozzolanic and blast furnace) or additives (bentonite, calcium chloride and sodium silicate) are mixed with grout to improve specific functions. Furthermore, there are recent studies (e.g., Yoshida 2012) indicating that a reuse of the spoil (whose disposal is costly and often problematic) could be convenient from an environmental point of view. In addition, as shown by various experimental evidence (e.g., Nikbakhtan and Osanloo 2009), technological details of the injection process or peculiarities of the site may lead to a large variability of mechanical properties and, most importantly, to a heterogeneity of the treated soil and, therefore, to a large scatter of such properties.

4.5.1 Material composition

The complex grout to soil interaction mechanism creates a body of jet-grouted material resting in place and an amount of spoil flowing to the ground surface. The composition of both materials depends primarily on the characteristics of the native soil, but non-negligible differences depend on the adopted jet grouting system. With regard to the former factor, noticeable attention has been devoted in the previous chapter (see Section 3.3), to the mechanisms induced on the different soil types, distinguishing the following three predominant phenomena:

1. Erosion of single soil particles induced by the dragging action of the grout (coarse soils)
2. Cutting of relatively large clods of soil (fine-grained soils)
3. Seepage of grout through soil pores (coarse interlocked soils)

As a consequence of these mechanisms, different textures are generally formed within the jet-grouted materials. Apart from the very peculiar effect observed on clean interlocked gravels (Figure 3.14), in which seepage is produced without changing the original fabric of the material, remoulding (erosion or clod cutting) is the ruling mechanism, producing the variety of structures shown in Figure 4.13. In coarse-grained soils, a texture formed by soil particles floating into a homogenous matrix of cement is obtained (Figure 4.13a). Such a deep modification of soil structure is not generally

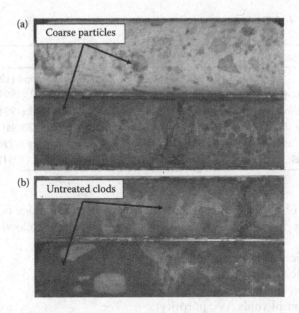

Figure 4.13 Samples cored from columns injected in coarse-grained (a) and fine-grained (b) soils.

obtained on finer soils (clays and silty clays). For them, the texture of the jet-grouted material is largely inhomogeneous, usually containing clods of uncemented material immersed in the grout matrix. In some cases (Figure 4.13b) the original soil structure may still be clearly visible even after treatment. A reduction in the dimension of clods may be obtained by multiple passes of the jet.

With regard to the influence of the injected fluids, no direct observation is available on the texture of a jet-grouted material created with different treatment systems. However, the systematically larger mechanical resistance of materials created with the single-fluid system in comparison with those produced by the double-fluid one suggests that, with the latter, bubbles of air remain trapped in the jet-grouted material.

4.5.2 Unit weight

The unit weight of the jet-grouted material is of paramount importance for those applications where the self-weight of the jet-grouted structures plays a relevant role. Sealing plugs created at the bottom of excavations or earth-retaining walls, in which a large part of the supporting function is provided by the self-weight, are the most enlightening examples. In addition, the

Table 4.10 Dry unit weight of jet-grouted material obtained with single-fluid system

Soil type	$\overline{\gamma}_{dc}(kN/m^3)$	$CV(\gamma_{dc})$	Reference
Clay	16.25	0.05	Botto and Capolupo (1989)
Silty clay	16.81	0.04	Xanthakos et al. (1994)
Silty sand	18.32	0.04	Xanthakos et al. (1994)
Sandy silt	17.89	0.08	Croce et al. (2004a)
Pyroclastic sand	13.89	0.07	Croce and Flora (1998)
Gravel	22.80	0.04	Mongiovì et al. (1991)

unit weight of the jet-grouted material can be seen as an index of treatment effectiveness. In principle, the unit weight depends on the following factors:

- Specific weight of the soil grains
- Specific weight of the hardened grout
- Relative amount of soil and grout
- Amount of voids (i.e., porosity)

A rule describing how the previously mentioned factors combine in different possible situations and for different treatment systems is not available, and thus, the unit weight must be experimentally determined.

As an example, the dry unit weights found by different authors for single-fluid jet grouting performed in different soils have been listed in Table 4.10. Each set of data has been treated as a statistical sample, and the computed mean value and variation coefficient have been reported in the table.

The comparison shows largely different dry unit weights, with the highest values $\left(\overline{\gamma}_{dc} = 22.80 \text{ kN/m}^3\right)$ pertaining to gravelly soils and the lowest ones $\left(\overline{\gamma}_{dc} = 13.89\right)$ pertaining to pyroclastic sands. In practice, it may be assumed that the jet-grouted material has a unit weight similar with that of the untreated soil when using single- and triple-fluid systems (Figure 4.14). With the double-fluid system, slightly lower unit weights are obtained because of the bubbles of injected air entrapped in the fresh mixture. This reduction is not observed in triple-fluid jet grouting because the air is injected together with the water from the upper nozzles and, therefore, it can be expelled by the subsequent remoulding of the soil caused by the grout jet.

4.5.3 Mechanical properties

Figure 4.15 reports the stress–strain response of a natural and of a jet-grouted sandy soil, showing the beneficial effect of treatment. Notwithstanding the positive effect of the confining stress on the behaviour of the untreated

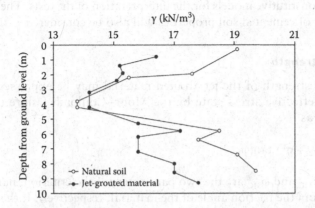

Figure 4.14 Unit weight of natural soil and jet-grouted material for single fluid jet grouting. (Modified from Tornaghi, R. and A. Perelli Cippo, Soil improvement by jet grouting for the solution of tunelling problems. *Proceedings of the Conference on Tunnelling*, Brighton, United Kingdom: pp. 265–275, 1985.)

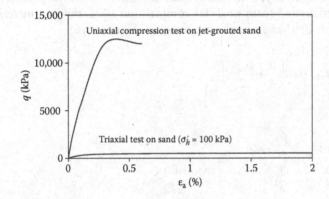

Figure 4.15 Typical strength increment provided by single-fluid jet grouting.

sand, the jet-grouted material is markedly stiffer and stronger than the natural soil. Because of the cementation, the jet-grouted material behaves as a soft rock.

In this section, attention will be devoted to the influence of the relevant factors (i.e., composition of the original soil, water content of the grout, curing time) on the strength and stiffness of jet-grouted material as observed from laboratory tests, whereas no attempt is made to introduce complex constitutive models for the interpretation of the tests. The inherent variability of cemented soil properties will also be considered.

4.5.3.1 Strength

The shear strength of the jet-grouted material may be expressed considering the effective stress state by the Mohr–Coulomb failure criterion, expressed as

$$\tau = c_{MC} + \sigma^{l} \cdot \tan(\varphi_{MC}) \tag{4.13}$$

in which c_{MC} and φ_{MC} are the two parameters of the criterion, namely, the cohesion and the friction angle of the material, respectively. If the effect of the stress state is neglected, the Tresca criterion may be adopted, for which the shear strength is expressed as a function of a single parameter in the form

$$\tau = c_{T} \tag{4.14}$$

The two criteria are obviously related. Since the Tresca parameter c_T is usually calculated from uniaxial compression tests, the parameters of the two models are linked by the relationship (Figure 4.16)

$$c_{T} = c_{MC} \cdot \tan\left(\frac{\pi}{4} + \frac{\varphi_{MC}}{2}\right) \tag{4.15}$$

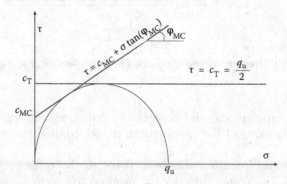

Figure 4.16 Failure criteria usually adopted for the jet-grouted material.

The Mohr–Coulomb criterion is familiar to geotechnical engineers and has the advantage of considering the beneficial effect of mass forces via the frictional term. However, it is required only when the frictional term is comparable to the cohesive one. This may happen for low values of the cohesion and/or particularly high confining stresses (i.e., very deep or highly confined treatments).

On the contrary, the Tresca criterion can be conveniently used for those applications for which the frictional term is negligible in comparison to the cohesive one, either because the confining stress is low or because the cohesion is high.

The results of several triaxial tests carried out on specimens of jet-grouted materials cored from a number of columns have been collected from three different case studies (Croce and Flora 1998; Fang et al. 2004; Bzówka 2009). Figure 4.17 shows the results in the s, t plane, where the two stress variables s and t are defined as

$$s = \frac{\sigma_1 + \sigma_3}{2} \tag{4.16a}$$

$$t = \frac{\sigma_1 - \sigma_3}{2} \tag{4.16b}$$

The linear trends shown in the figure by all results confirm that, at least at the analysed stress levels, the assumption of a linear failure envelope is reasonable.

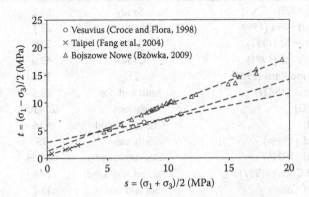

Figure 4.17 Failure envelopes of jet-grouted material from three case studies.

The Mohr–Coulomb failure parameters collected from a number of case studies (including the previously considered ones) are reported in Table 4.11. From the comparative analysis of all the data, it clearly appears that no correlation exists between the friction angles of the jet-grouted material and of the original soils (not reported in the table). For instance, the two cases reported by Mongiovì et al. (1991), in which jet grouting is performed on gravelly soils, present friction angles of the jet-grouted material lower than the one obtained on a treated sandy soil by Bzówka (2009), whereas the friction angles of treated clayey soils range between 28.5° and 44°, the latter value being unrealistic for untreated fine-grained materials.

The lack of relationship between the untreated and the treated soil friction angle seems to be a logical consequence of the fact that the friction angle of a granular material is primarily related to grading, density and fabric, and all these characteristics are deeply modified by erosion, mixing and washing operated during jet grouting. As a consequence, the only way to estimate the two parameters c and φ would be by carrying out triaxial tests, which is not common in the case of jet-grouted materials mostly because triaxial apparatuses able to apply high or very high confining stresses are required.

In practice, considering the relatively high cohesion and the comparatively limited confining stresses usually acting on jet-grouted structures, the jet-grouted material is more frequently considered as a low-quality concrete, and the simple Tresca criterion is adopted, characterising the failure envelope with the uniaxial compressive strength q_u.

Table 4.11 Summary of Mohr–Coulomb parameters from various case studies

Reference	Soil type	φ_T (°)	c_T (MPa)
Bzòwka (2009)	Sandy	58.2	2.3
Croce and Flora (1998)	Silty sand	26.1	3.2
Mongiovì et al. (1991)	Gravel	52.0	2.1
Mongiovì et al. (1991)	Gravel	42.0	0.3
Mitchell and Katti (1981)	Clay	39.5	0.58
Yahiro et al. (1982)	Sand and clay	28.5	0.4–1.0
Miki (1982)	Various	20–30	0.7–1.0
Yu (1994)	Clay–silty sand	40.6	1.1
Fang et al. (1994a)	Silty sand	35	4.2
Fang et al. (1994b)	Clay–silty sand	40–44	4.2
Fang and Chung (1997)	Clay and silty sand	38.6	0.8
Fang et al. (2004)	Silt and sand	38.7	0.7
Nikbakhtan and Osanloo (2009)	Clay and sand	42–49	0.4–0.8
	Clay and sand	25	0.77

Indicative ranges of q_u for different soils are given in Figure 4.18 as a function of the amount of injected cement (Fiorotto 2000). The figure shows a very large resistance for jet grouting in coarse-grained soils, with values slightly lower than typical concrete for high cement dosages. The values of q_u largely reduce, even for relatively high cement dosages, in fine-grained soils. This is a well-known feature of concrete, whose mechanical properties strongly depend on the quality of the aggregate. In jet grouting, the poor mechanical effect of cementation in clays and silty clays may be amplified by the existence of clods of uncemented material (see Figure 4.13).

Xanthakos et al. (1994) give similar indications on the q_u of jet-grouted materials, suggesting values of the uniaxial compressive strength between 1.5 and 10 MPa for fine-grained soils and between 10 and 30 MPa for coarse-grained soils.

More detailed experimental evidence shows that, similar with what is typically seen on natural soft rocks, the uniaxial compressive strength q_u is strongly related to the jet-grouted material's dry unit weight γ_{dc}, the relation depending on native soil gradation (Figure 4.19). Croce and Flora (1998) and Croce et al. (2004a) experimentally observed that the increase in both γ_{dc} and q_u is somehow related to the increase in depth.

The uniaxial compressive strength q_u depends on the treatment system (single, double or triple fluid) as well as on the cement–water ratio. As previously mentioned, because double-fluid jet grouting is expected to give the lowest possible weight of the cemented material, it is also expected to generate the lowest possible compressive strength. This is confirmed by the experimental results reported in Figure 4.20 (van der Stoel 2001).

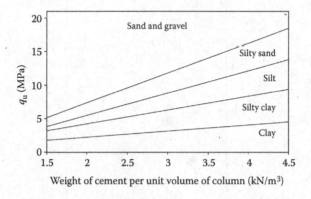

Figure 4.18 Indicative ranges of uniaxial compressive strength for different soil types and variable injected amounts of cement. (Modified from Fiorotto, R., *Improvement of the Mechanical Characteristics of Soils by Jet Grouting*, personal communication, 2000.)

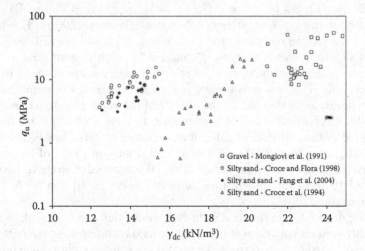

Figure 4.19 Uniaxial compressive strength versus dry unit weight of the jet-grouted material.

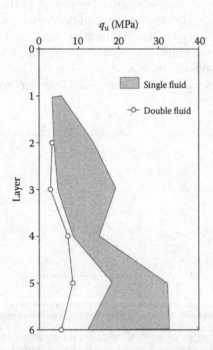

Figure 4.20 Uniaxial compression strength versus depth for single- and double-fluid jet grouting. (Modified from van der Stoel, A. E. C., *Grouting for Pile Foundation Improvement*: PhD thesis, Delft University Press, 217 pp., 2001.)

The effect of the cement–water ratio on the behaviour of the consolidated soil is consistent with concrete engineering. In particular, increasing the cement–water ratio of the grout generally leads to an increase in strength and to an increase in hardening time (but this latter effect can be shadowed by the use of hardening accelerating additives). Figure 4.21 shows such evidence for an alluvial sandy soil (Arroyo et al. 2007). The figure also shows that the differences in uniaxial compressive strength tend to reduce as the depth increases, thus suggesting a possible difference in jet-grouted material composition with depth. Data reported by Kutzner (1996) show reductions of the compressive strength of up to 50% when reducing the cement–water ratio of the injected grout from 1.5 to 1.

A final, relevant aspect related with the strength of jet-grouted material, which must be carefully considered when conceiving the construction sequence of a group of columns, is the influence of curing time. For jet-grouted material, the development of strength is in all similar with that of concrete, with fast gains in the first days after jetting and with gradients progressively reducing with time. Figure 4.22 reports the experimental results from the well-documented case of the new Barcelona metroline (Arroyo et al. 2007), showing that most of the strength is developed in the first seven days (from 60% to 90% of the resistance measured after 28 days).

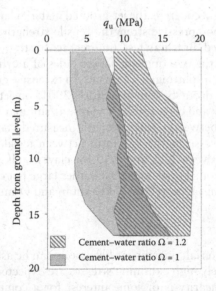

Figure 4.21 Uniaxial compression strength versus depth for single-fluid jet grouting performed in sandy soil, with different cement–water ratios by weight. (From Arroyo, M. et al., *Informes Sobre Tratamientos de Jet Grouting*. ADIF LAV Madrid-Barcelona-Francia, Tramo Torrasa-Sants. Report of the Universidad Politecnicha de Catalunya, p. 110 [in Spanish], 2007.)

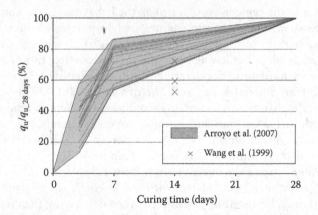

Figure 4.22 Uniaxial compression strength versus time for jet grouting performed in sandy soil and in marine clays. (From Arroyo, M. et al., *Informes Sobre Tratamientos de Jet Grouting*. ADIF LAV Madrid-Barcelona-Francia, Tramo Torrasa-Sants. Report of the Universidad Politecnicha de Catalunya, p. 110 [in Spanish], 2007; Wang, J. G. et al., Effect of different jet grouting installations on neighbouring structures, In Leung C. F., Tan S. A. and Phoon K. K., eds., *Field Measurements in Geomechanics*: Rotterdam, Netherlands: Balkema: pp. 511–516, 1999.)

Like soft rocks and concrete, the jet-grouted material has a tensile strength much lower than compressive strength. Tensile strength is typically evaluated via the so-called 'Brazilian test' (indirect tensile test), that is, by loading in compression on the two opposite lateral sides of a cylindrical specimen, reaching failure via a splitting mechanism. An extensive experimental investigation was carried out by van der Stoel (2001) on jet-grouted material coming from sandy and clayey soils. The results summarised in Figure 4.23 show the relation between tensile strength measured via the Brazilian test and the compressive strength. The ratio between tensile and compressive strength ranges between 1/15 and 1/10 for clayey soils and between 1/10 and 1/5 for sandy soil. These ranges are rather large but consistent with the experimental findings reported by Nikbakhtan and Osanloo (2009).

4.5.3.2 Stiffness

For jet-grouted material, quasilinear behaviour can be assumed before failure, unless extremely high confining stresses are expected.

However, nonlinearity is of some interest for a complete characterisation of the stress strain behaviour of the cemented material at very small strains. Figure 4.24 reports an experimental comparison carried out in uniaxial compression tests between the small strain Young modulus (E_0) and the tangent modulus corresponding to half of the uniaxial compressive

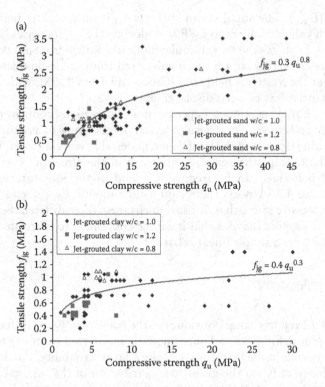

Figure 4.23 Tensile versus compressive strength for jet-grouted materials. (a) Sandy soil, (b) clayey soil. (Modified from van der Stoel, A. E. C., *Grouting for Pile Foundation Improvement*: PhD thesis, Delft University Press, 217 pp., 2001.)

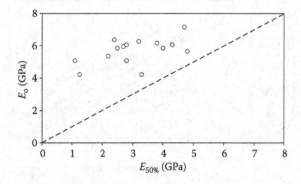

Figure 4.24 Large strain Young modulus $E_{g,50\%}$ versus small strain Young modulus $E_{g,0}$ of jet grouted material. (Modified from Fang, Y. S. et al., Properties of soilcrete stabilized with jet grouting. *Proceedings of the 14th International Offshore and Polar Engineering Conference*, Toulon, France, May 23–28, 2004: pp. 696–702, 2004.)

strength ($E_{50\%}$). The small strain stiffness E_0 is only slightly higher than $E_{50\%}$, with values of the ratio $E_0/E_{50\%}$ reducing as $E_{50\%}$ increases.

Figure 4.25 shows the secant Young moduli obtained from uniaxial compression tests carried out on specimens cored from single-fluid jet-grouted columns at the Vesuvius field trial (Croce and Flora 1998), with accurate local measurements of axial displacements.

The previously mentioned experimental observations confirm that it is reasonable and convenient from a practical point of view to assume that the stress–strain response of the jet-grouted material at working levels is linear.

The relationship between the stiffness and the strength of artificially cemented materials can be frequently found in the literature (e.g., Bell 1993). Figure 4.26 reports the secant Young moduli $E_{50\%}$ versus the uniaxial compressive strength q_u for specimens from three different field trials. The results support the idea that it can be convenient to correlate stiffness and strength via a simple linear relation:

$$E_{50\%} = \beta_E \cdot q_u \tag{4.17}$$

Table 4.12 reports ranges of values of the coefficient β_E for different soils and different definitions of Young moduli, as inferred from experimental studies reported in the literature. Despite a large variability, the data show a dependency of β_E on the grain size distribution of the original soil: the highest values of β_E (~1000–1200) are obtained in coarse soils (gravel and sandy gravel), whereas the lowest values (~100–200) pertain to jet grouting carried out in finer soils (from clay to silty sand).

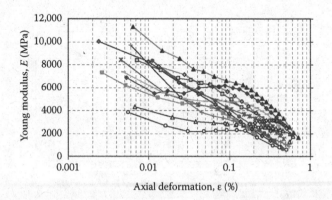

Figure 4.25 Young modulus decay during uniaxial compression tests on jet-grouted material strength. (Modified from Croce, P., A. Flora and G. Modoni, Experimental investigation of jet grouting. *Proceedings of the ASCE Conference '2001 a GeoOdissey'*: Blacksburg, VA: Virginia Tech: pp. 245–259, 2001.)

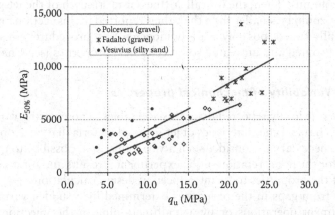

Figure 4.26 Young modulus versus uniaxial compression strength of jet-grouted material from different case studies. (Modified from Modoni, G. and J. Bzówka, *ASCE Journal of Geotechnical and Geoenvironmental Engineering* 138(12): pp. 1442–1454, 2012.)

Table 4.12 Relationship between young moduli and uniaxial compressive strength of jet-grouted material from the literature

Reference	Definition of E	Soil type	β_E
Mongiovì et al. (1991)	Tangent unspecified	Gravel	280–1000
Lunardi (1992)	Secant at 40% q_u	Gravel and sand	500–1200
Nanni et al. (2004)	Tangent unspecified	Gravel and sand	440–1000
Croce et al. (1994)	Tangent unspecified	Sandy gravel	210–670
Croce and Flora (1998)	Secant at $\varepsilon_a = 0.01\%$	Silty sand	220–700
Nanni et al. (2004)	Tangent unspecified	Silty sand	330–830
Fang et al. (2004)	Tangent at 50% of the failure stress	Silty sand	300–750
Fang et al. (2004)	Tangent at 50% of the failure stress	Silty sand, silty clay	100–300
Lunardi (1992)	Secant at 40% q_u	Silt and clay	200–500

It is finally worth noticing that all the considerations reported in this section pertain to the behaviour of the specimens of jet grouting tested in the laboratory. The variability of the properties within the column will be discussed in the next section. What is important to stress here is that specimens tested in the laboratory are those that can be retrieved from coring and so represent the best parts of the columns. In jet-grouted structures, not all the soil is always completely grouted, either because of defects or because geometrical grids are purposely assigned to leave untreated soil among columns. Then, the overall stiffness and strength of the jet-grouted structure could be smaller than the values reported in this section and must be carefully tuned, possibly via a homogenisation procedure, considering both jet-grouted and untreated soil mechanical properties (see Chapter 6).

4.5.3.3 Variability of mechanical properties

In all the previously reported diagrams and tables, a noteworthy variability of the mechanical characteristics of jet-grouted material has been observed. It is then necessary to consider such a variability and, possibly, to quantify it, both for the interpretation of the experimental results and for a safe definition of the properties to be introduced in design calculation.

Apart from gaps in the properties determined by a sudden variation of the treatment operations or by a malfunctioning of the injection equipment, which can be, nowadays, recognised and corrected with the aid of automatic systems recording the treatment parameters (see Chapter 8), the observed variability of mechanical properties is basically a product of the variable composition of the cemented material. In fact, the properties of jet-grouted materials depend on the jet–soil interaction mechanism (mixing, cutting of clods, replacement, seepage), which, in turn, depends on both native soil properties and jet grouting procedure (in terms of system and treatment parameters). The inherent spatial variability of the natural soil composition and mechanical properties affects the interaction mechanism and produces, for what has been shown before, variable properties within the cemented soil body.

Before quantifying this aspect, it is worth considering that the phenomenon is largely dependent on the geometrical scale. In fact, although the mathematical definition of resistance and stiffness for a continuum is referred to a single point, the experimental quantification is provided on specimens of small but finite dimensions. In design calculations, these properties are extrapolated and referred to the cemented soil masses of much larger dimensions. One possibility to include the variability of the properties in design calculations could be theoretically represented by a quantification of the spatial correlation, that is, by relating the variability of properties to the position in which they were measured. However, this procedure is usually time consuming because a large number of samples should be cored and their position should be fixed with accuracy, and can be worth carrying out only in the case of massive treatments performed in highly heterogeneous soils (e.g., superposed thin layers of different materials).

In most cases, the jet-grouted structure has a relatively limited extension. As a consequence, it can be difficult to relate variability to position, and a single value of the mechanical property is typically assumed. This value can be considered constant within the whole structure or probabilistically variable among the different sections (see Section 6.4.3). Then, the problem arises to quantify a mean value and a scatter for the jet-grouted structure starting from the values experimentally measured on samples. The test results can be assumed as a statistical sample, and probabilistic models can be introduced to simulate the observed variability.

Given a mean μ and a standard deviation SD of the values measured on specimens, the mean and standard deviation for the jet-grouted structure can be calculated as follows:

$$\mu(\text{structure}) = \mu(\text{samples}) \tag{4.18}$$

$$SD(\text{structure}) = \frac{1}{\sqrt{a}} SD(\text{samples}) \tag{4.19}$$

in which a is the ratio between the area of the cross-section of the structure and that of the tested specimen. Considering the typical diameters of the jet-grouted columns (say, 1–3 m) and of the cored specimens (0.08–0.10 m), the parameter a of Equation 4.19 ranges from 100 to 1500. Therefore, the variability of the mechanical properties of a column is expected to be from 1/10 to 1/40 of the variability inferred from the laboratory results. If the reduction of scatter caused by an increase in scale is neglected at the design stage, the variability assigned to the properties of the columns is overestimated.

The variability of the uniaxial compressive strength of specimens cored from jet-grouted columns can be seen from Figure 4.27, in which the frequency distributions reported in the literature for six different case studies are reported, each referring to a different soil. In the plots, each q_u value has been divided by the mean value of its series. Consistently with all the indications of literature, plots obtained for clayey, sandy and gravelly materials show an asymmetric bell-shaped distribution.

The computed coefficients of variation CV, also reported in the figures, show that the scattering of data for sandy and gravelly soils is lower ($0.15 \leq CV \leq 0.47$) than the one pertaining to finer soils ($0.48 \leq CV \leq 0.75$). These results can be explained with the physical and mechanical properties of jet-grouted soils. As extensively reported in the first part of this chapter, soil erosion by jet grouting is far more effective in granular than in fine-grained soils, where the mutual adhesion between particles makes disaggregation more difficult. As a consequence, jet-grouted materials are usually more homogenous if jet grouting is performed in granular soils. On the contrary, jet-grouted materials obtained in finer soils may include lumps of untreated soils and thus have a rather variable degree of cementation. Furthermore, there is an obvious physical lower limit to the

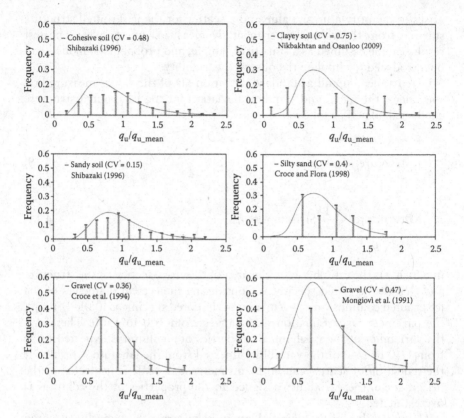

Figure 4.27 Frequency distribution of the uniaxial compression strength from six different case studies.

value of the uniaxial compressive strength (which is zero, that is, it cannot be negative), and therefore, a symmetric distribution is possible only
when the mean value is not too low and the scatter is not too large. Since
jet-grouted materials obtained in fine-grained soils have a lower value of
q_{u_mean} and a higher scatter of results, it must be expected that they show a
higher asymmetry.

Since only six experimental data sets are reported in Figure 4.27, the values of CV must be considered as qualitative indications of the behaviour of
jet-grouted materials and cannot be considered as representing all possible
situations. For instance, in a given soil, the effectiveness of erosion strongly
depends on the specific energy of jet grouting and, as a consequence, the
scatter of physical and mechanical properties should be expected to reduce
as jet energy increases.

It is interesting to note that all plots reported in Figure 4.27 can be interpreted by a log-normal probabilistic model. In fact, the continuous curves in the figures represent the values of the product $f(x) \cdot \Delta x$, where $x = q_u/q_{u\_mean}$ and $f(x)$ is expressed as

$$f(x) = \frac{1}{x \cdot \sigma \cdot \sqrt{2\pi}} \exp\left(-\frac{\ln x - \mu}{2\sigma^2}\right) \tag{4.20}$$

The mean μ and variance σ^2 of the logarithmic distribution can be calculated starting from the mean $\bar{x}$ and the coefficient of variation $CV(x)$ of the samples with the following correlations:

$$\mu = \ln(\bar{x}) - \frac{1}{2}\sigma \tag{4.21}$$

$$\sigma^2 = \ln[1 - CV^2(x)] \tag{4.22}$$

Since it has been previously shown that there is a link between jet-grouted material strength and stiffness (Equation 4.17), similar considerations on the variability of values can be drawn for Young moduli. The statistical analysis of the Young moduli reported in Figure 4.28 (Katzenbach et al. 2001) is just an example to confirm the expected asymmetric distribution of the experimental values, with the lower values more probable and the log-normal probabilistic model sufficiently accurate.

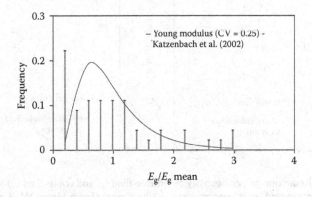

Figure 4.28 Frequency distribution of the Young moduli. (Modified from Katzenbach, R. et al., Jet grouting: Chance of risk assessment based on probabilistic methods. In Durgunoglu, H. T. ed., *Proceedings of the 15th ICSMFE*, Istanbul, Turkey: pp. 1763–1766, August 27–31, 2001.)

4.5.4 Permeability

The permeability of a jet-grouted material is usually very low, being the material generally composed of soil particles, either disaggregated or in lumps, immersed in a fine matrix of hardened cement. In practice, the Darcy coefficient of permeability k of the treated soil (pertaining to a homogenous, completely treated volume of soil or at the laboratory scale) has values between 10^{-7} and 10^{-9} m/s. An exception is represented by the case of clean, highly interlocked gravels, for which it may happen that jet grouting is mostly effective by permeation, and the injected grout is not able to fully saturate the pores (see Figure 3.15), thus resulting into higher coefficients of permeability.

On site, the overall permeability of groups of jet-grouted columns is mostly governed by defects. Such defects may play a minor role on the overall seepage processes if jet grouting is carried out in fine-grained soils, whereas they may become relevant in granular soils because of the localised flows through the untreated soil portions. In the latter case, the overall permeability on site may be much higher than the one measured in the laboratory on small specimens. One of the few studies reported in the literature

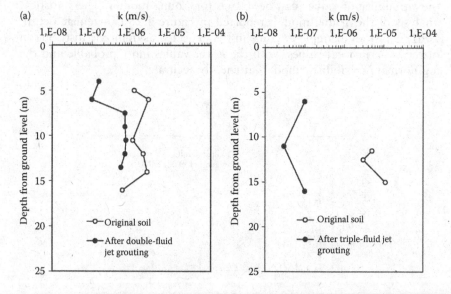

Figure 4.29 Reduction of permeability by double-fluid (a) and triple-fluid (b) jet grouting measured with Lugeon tests in silty sands. (From Hong, W. P. et al., Case study on ground improvement by high-pressure jet grouting. *Proceedings of the 12th International Conference on Offshore and Polar Engineering,* International Society of Offshore and Polar Engineers, Kitakyushu, Japan, May 26–31, 2002: pp. 610–615, 2002.)

on the overall permeability of jet-grouted structures on site is that by Hong et al. (2002), who performed a series of Lugeon tests on a sandy gravel treated with double- and triple-fluid jet grouting techniques. The results, summarised in Figure 4.29, show that treatments were able to reduce the permeability of the original soil by one or two orders of magnitude. The figure also shows that the triple-fluid system produces values slightly lower than those obtained with the double-fluid system.

Chapter 5

Jet-grouted structures

5.1 ELEMENTS AND STRUCTURES

The success of jet grouting is probably caused by the unique advantage of creating in the subsurface consolidated elements of various shapes and sizes, provided with good mechanical properties and reduced permeability. Jet grouting is also very attractive because the treatments can also be performed in difficult operating conditions, working in confined spaces or in places difficult to reach with other means.

Complex elements may be obtained by conveniently assembling more columns and by resorting to various possible constructive measures, for example by varying the inclination of the treatments or by interrupting grouting in some sections of drilling. The jet-grouted elements can be also reinforced with the insertion of metal or fibreglass bars or tubes, which can provide flexural and tensile resistance if needed (see Section 6.4.4).

In any case, considering the wide variety of geometries that can be obtained by jet grouting treatments, it is possible to classify the consolidated elements based on both their shape and their function. In particular, with reference to the first aspect, it is possible to distinguish three main categories, namely, one-dimensional, two-dimensional and three-dimensional elements.

The most common one-dimensional elements consist of individual isolated columns, the length of which is much larger than the diameter. A typical jet-grouted structure made by such one-dimensional elements is represented by so-called 'reinforced foundations', in which the arrangement of the jet-grouted columns follows the geometry of the traditional pile rafts and may thus be called 'jet-grouted raft'. In this case, the jet-grouted columns perform the function of increasing the allowable load and reducing the settlements of the overlying structure.

On the contrary, when the treatments are placed side by side, it is possible to obtain two- to three-dimensional consolidated elements of various types (Figure 5.1). In fact, if the treatment axes are close enough, the jet columns

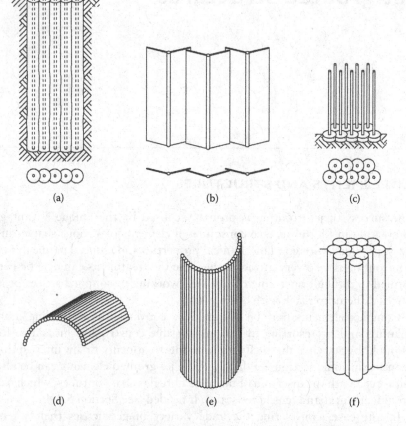

Figure 5.1 Two- and three-dimensional elements: (a and b) jet-grouted diaphragms; (c) jet-grouted slabs; (d) jet-grouted canopy; (e) jet-grouted shaft; (f) jet-grouted block.

become interpenetrated along single or multiple lines to form elements of planar, cylindrical or conical shape, according to the design needs.

Vertical planar elements, obtained by joining long jet-grouted columns (Figure 5.1a), are frequently used to create jet-grouted diaphragms that can be used as soil-retaining structures and/or hydraulic cutoffs. The latter function can be also obtained by partially overlapping contiguous V-shaped elements (Figure 5.1b). Horizontal planar elements, commonly called 'jet-grouted slabs', can be used to create horizontal water barriers at the bottom of excavations (Figure 5.1c). These elements are obtained by limiting the grouting length to a relatively short interval.

The jet columns are also sometimes arranged to form a frustum of cone (Figure 5.1d), which can be used as a canopy for the provisional support of tunnelling, or in a cylindrical pattern used to excavate large diameter shafts (Figure 5.1e).

Massive three-dimensional elements, usually named 'jet-grouted blocks', are obtained by performing jet grouting treatments at very reduced mutual distance to consolidate large volumes of soils (Figure 5.1f). In fact, by following such procedure, a large block of consolidated material can be obtained with the aim of improving the bearing capacity of a foundation to make a large gravity earth retaining wall or to create a bottom-water sealing plug.

These basic elements can be combined in multiple arrangements to form various 'jet-grouted structures' designed to fulfil specific functions.

With regard to the possible design purposes, treatments may be performed not only for the accomplishment of a new construction but also for the improvement or modification of an already existing one. Jet grouting structures may have permanent (i.e., long-lasting) functions or provisional (i.e., temporary) functions.

In this chapter, the typical applications of jet grouting will be considered, grouping them as follows:

- Foundations
- Retaining structures
- Water barriers
- Tunnels
- Other applications

For each application, different kinds of solutions will be presented, highlighting their respective advantages and limitations.

Design hints and examples on some of these jet-grouted structures are reported in Chapter 7, based on the general design considerations discussed in Chapter 6.

5.2 FOUNDATIONS

Jet grouting is frequently used for soil improvement under concrete foundations, such as isolated footings, beams or slabs, with the double purpose of increasing the bearing capacity and reducing the settlements of the overlying structure. In general, the plan extension of the treatment mirrors that of the overlying structure, whereas the other geometrical features to be defined are the length, the diameter and the spacing of the jet columns.

Two different types of solutions should, however, be distinguished. The first one is obtained when the centre-to-centre distance (spacing) of the drilling axes is greater than the diameter of the columns, and so, each treatment gives rise to a slender cylindrical body (one-dimensional element), separated from the others. In such first case, the arrangement of the columns is then similar to that of a pile raft, and the resulting foundation could thus be named 'jet-grouted raft' (Figure 5.2a). Jet-grouted rafts are frequently chosen as an alternative to conventional pile rafts when the subsoil characteristics are unfavourable for the installation of drilled or driven piles, as happens, for instance, when there are large rock masses interspersed in the soil matrix. In general, jet-grouted rafts are quite effective for settlement reduction, but in many cases, they do not provide adequate resistance to horizontal actions because of their low flexural and shear strength. Even when steel bars or tubes are inserted into the jet columns, the flexural strength remains relatively low. To overcome this important limitation, some of the columns may be properly inclined to take up the horizontal load component as shown in Figure 5.2b (Falcao et al. 2001).

If the load eccentricity is very high, it may be more convenient to create massive reinforcements, obtained when the centre-to-centre distance of drillings is smaller than or close to the diameter of the columns. In this case, the treatments produce a single large block of jet-grouted material, having the shape of a large diameter cylinder or a wide parallelepiped (Figure 5.2c). The block can be reinforced by a relevant number of steel bars or tubes, extending upward into the superimposed reinforced concrete slab, thus providing effective resistance to tensile and shear stresses.

One of the first projects of this kind reported in the literature concerns a highway viaduct built in Italy in the 1990s (Croce et al. 1990). This viaduct is composed of two independent structures, one for each driving direction, that have common foundations. The latter are located along a steep slope on a thick deposit of coarse-grained soils (Figure 5.3). Each foundation was built

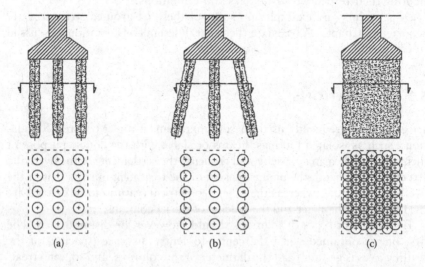

Figure 5.2 Typical jet-grouted foundations: (a and b) jet-grouted raft; (c) jet-grouted block.

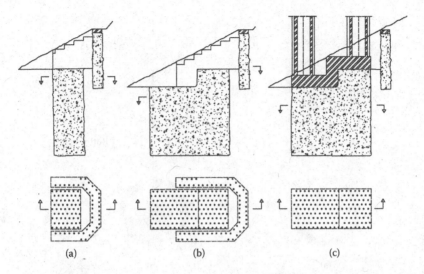

Figure 5.3 Construction sequence of a foundation made by jet-grouted block: Fadalto Viaduct case. (a) First construction step: earth retaining jet-grouted structure, top excavation and uphill jet-grouted block. (b) Second construction step: downhill excavation and completion of jet grouted block. (c) Third construction step: foundation cap by reinforced concrete. (Adapted from Croce, P., et al., *Rassegna dei Lavori Pubblici*, November: pp. 249–260 [in Italian], 1990)

by making a large jet-grouted block that consists of 166 vertical treatments performed with a triangular grid array, with spacing among the columns of 1.10 m. Given the inclination of the slope, it was necessary to proceed step by step, moving downward along the slope. In fact, a provisional supporting structure, made by a semicylindrical jet grouting element, was first provided uphill. The jet block foundation was then created by two subsequent series of treatments, working from two staggered levels, and was finally reinforced by means of steel tubes placed into the jet-grouted material by subsequent drilling and grouting.

Another peculiar example of massive reinforcement for a foundation to be built in difficult subsoil conditions is reported by Langhorst (2012). In this case, the jet grouting reinforcement was necessary to increase the bearing capacity and reduce settlements of pillars and abutments of a bridge located in a polder characterised by a significant presence of clayey materials. The foundation had to transfer loads to deeper strata without damaging a polyvinyl carbonate (PVC) foil located below the foundation to seal the polder. In this example, special care has been posed to the selection of the jet grouting parameters (in particular, the type of cement and its ratio to water) to ensure adequate resistance and stiffness to the grouted material.

A different kind of massive foundation can be obtained by digging a large shaft that is then filled with concrete (e.g., Balossi Rastelli and Profeta 1985).

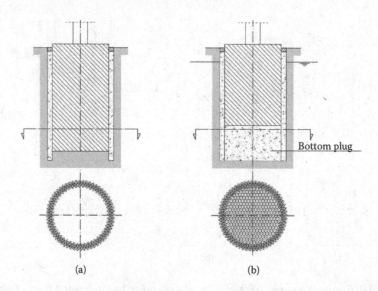

Figure 5.4 Typical jet-grouted shaft foundations: (a) above water level; (b) below water level.

In this case, jet grouting is frequently used to ensure the stability of shaft excavation. To this aim, the jet columns are arranged along the shaft contour, in single or double rows, so as to achieve a consolidated two-dimensional element of cylindrical shape (Figure 5.4a). If the cylindrical element is effectively continuous, excavation of the shaft may be performed without introducing additional support, given that the consolidated element behaves as a sequence of rings mostly subjected to horizontal compressive stresses (see Section 7.3.2). In practice, however, it is recommended to insert vertical steel reinforcements along the jet columns and/or circular steel ribs to provide some resistance at possible weak spots. For shafts excavated below the water level, it is also recommended to provide additional support and waterproofing by means of shotcrete because local treatment defects could result in uncontrolled water inflow that may compromise safety during excavation.

Sometimes, these large shaft foundations are made by two superimposed parts. The upper part is obtained by shaft excavation and subsequent concreting as it was previously described, whereas the lower one consists of a cylindrical jet block (Figure 5.4b). This latter solution has the advantage of limiting the excavation depth. It is based on the consideration that, because of the increase with depth of the self-weight of the block, the load eccentricity decreases with depth, and the tensile stresses tend to vanish accordingly (Croce et al. 2006). Therefore, there is no need to extend the reinforced concrete part of the foundation below a given depth. Moreover, if the shaft is to be excavated below the water level, the jet block becomes also a sort of bottom plug, providing the necessary water tightness during construction

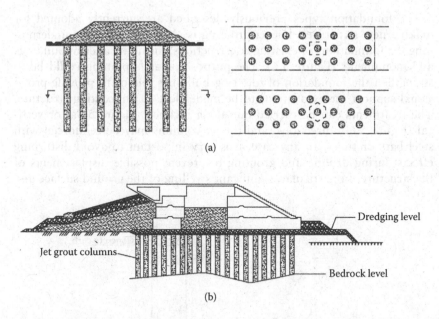

(a)

(b)

Figure 5.5 Foundation reinforcement with isolated jet-grouted columns: (a) embankment foundation. (b) Offshore foundation. (Modified from Croce, P. and G. Modoni, *Rivista Italiana di Geotecnica* 44(3): pp. 30–5 [in Italian], 2010; Durgunoglu, T. et al., Offshore jet grouting: A case study. *Proceedings of the ISSMGE–TC211 International Symposium on Ground Improvement*, Brussels, Belgium, May 31–June 1, 2012: pp. 225–234, 2012.)

(Figure 5.4b). Clearly, in such latter case, the jet block must be created from the ground level before starting excavation.

Two cases of foundation reinforcement with isolated columns are reported in Figure 5.5. In the former example (Figure 5.5a), jet grouting is created prior the construction of embankments to reduce the settlements induced by the compression of soft soils (Croce and Modoni 2010). Many similar applications are reported in the literature (e.g., de Paoli et al. 1989; Laguzzi and Pedemonte 1991; Alzamora et al. 2000; Pinto et al. 2012). Particular care has to be taken with the connection of columns with the overlying embankment to avoid excessive stress concentration and subsequent punching. Possible countermeasures consist of reducing the distance among columns (with spacing not exceeding 2–2.5 times the diameter) or by equipping the top of the columns with reinforced concrete caps or with multiple geogrid layers.

In the second example (Figure 5.5b), a dense grid of isolated jet-grouted columns was created to support an artificial island (the Sculpture Plaza at Doha, Qatar). The jet-grouted columns, designed to have a 0.80-m diameter and different lengths to reach the bedrock, were spaced at variable centre-to-centre distances (1.0–2.0 m) depending on the load transferred by the overlaying structure.

The foundation types previously described are currently adopted for making new structures, but jet grouting is also often used for underpinning the foundations of existing constructions, including ancient buildings and monuments (Figure 5.6). The purpose of underpinning could be to strengthen the foundation of tottering buildings and/or to provide provisional support for excavations to be made close to an existing structure. The treatments can be carried out along inclined (Figure 5.6a) or vertical (Figure 5.6b) directions, and the jet columns may be reinforced with steel bars or tubes. In any case, it is very important to avoid disturbing effects during drilling and grouting to prevent possible displacements of the structure. In particular, significant swelling of the ground surface and

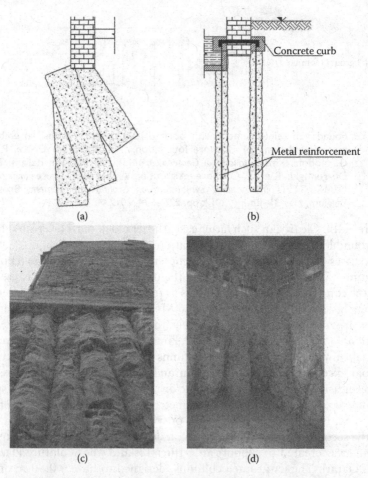

Figure 5.6 Typical solutions for jet-grouted underpinning: (a) unreinforced; (b) reinforced; (c and d) photograph of underpinning columns after excavation. (Courtesy of K. Wanik.)

consequent lifting of the existing foundations may occur if the outflow of the spoil is temporarily obstructed by borehole clogging or in the case of excessively high treatment energies. The opposite problem (i.e., settlement of the structure) can occur when jet grouting is performed in unsaturated, collapsible soils.

In most underpinning projects, it is also necessary to provide adequate connections between the jet columns and the existing foundations. The latter requirement may be accomplished by means of connecting beams made of steel or reinforced concrete (Figure 5.6b). Different examples of underpinning

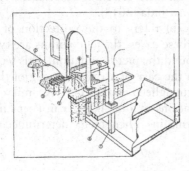

(1) Initial condition
(2) Execution of provisional jet grouting columns
(3) Insertion of steel beams
(4) Concrete casting
(5) Excavation between provisional columns
(6) Construction of the supporting beam
(7) Prolongation of columns
(8) Provisional support

(a)

Reinforced concrete

(b)

Figure 5.7 (a) Example of jet grouting application for the underpinning of old buildings. (b) The scouring protection of a bridge pier. (Modified from Garassino, A. L., Uso delle colonne in terra stabilizzata con jet grouting come elemento provvisionale. *Proceedings of the 15th National Geotechnical Conference 2*, Spoleto, Italy, Associazione Geotecnica Italiana: pp. 155–161 [in Italian], 1983.)

works are reported in the literature (Garassino 1983; Ichihashi et al. 1992; Popa 2001).

Two remarkable examples of jet grouting application for the improvement of existing structures are reported in Figure 5.7. In the former example (Figure 5.7a), regarding the underpinning of a historic building near Parma (Italy), the jet grouting columns (connected at their top by a reinforced concrete beam) served as a temporary support of the upper structure. An underground prolongation of upper columns was then built with reinforced concrete to create a lower level. A similar application, accomplished at the border of an ancient building also with earth-retaining functions, is reported by Burke (2007a,b) and Burke and Yoshida (2013).

The second example (Figure 5.7b) refers to the protection of bridge piers from hydraulic scouring. In these cases, the jet columns are typically arranged to form a sort of ring around the pier foundation. This technique has been adopted, for example, in the case of a river crossing of the road Bazoul Constantine in Algeria (Fruguglietti et al. 1989). A similar operation was carried out on the foundations of a breakwater in Port Elizabeth in South Africa to prevent undermining phenomena determined by the waves (Parry Davies et al. 1992).

5.3 RETAINING STRUCTURES

Jet grouting treatments are frequently used to create provisional, or, sometimes, permanent, retaining structures. To this aim, it is necessary that the columns are partially overlapped to form continuous two- or three-dimensional jet-grouted elements.

In excavations with circular plans, treatments are configured so as to produce a shell structure, having a cylindrical shape, similar to the one previously described in the case of shaft foundations (see Figure 5.4). On the contrary, for large-size excavations or with linear plan arrangement, the jet-grouted columns are positioned along one or more parallel lines to form long vertical diaphragms. The stresses acting on the consolidated elements are clearly very different for each of the two previously outlined cases.

In fact, a cylindrical supporting element provides an arching effect in the horizontal direction, and thus, the consolidated material is predominantly subjected to compressive stresses that are particularly suited to the mechanical characteristics of the consolidated material (see Chapter 4). The effectiveness of jet grouting, for this kind of earth-retaining structure, is therefore assured as long as the columns are sufficiently interpenetrated to ensure the continuity of the jet-grouted element with adequate thicknesses, and their plan arrangement is able to generate compressive stresses in the whole jet-grouted volume. Because the two conditions may be negatively affected by deviations of columns from their prescribed direction, great attention should be paid to the control of verticality when performing treatments.

One of the first cases of a jet-grouted retaining structure having a cylindrical shape is reported by Balossi Restelli et al. (1986). The treatment was carried out for provisional support to create a shaft having a 12.4-m diameter and a 22-m depth (Figure 5.8), which was needed to provide access to the train tunnel of a metroline in Milan. The work was executed in a particularly difficult context, sensitive to the presence of buildings rising at a very short distance from the shaft. To carry out the excavation, it was necessary to contain the settlements of the surrounding buildings within allowable values. To this aim, a cylindrical shell was created prior to the excavation with interpenetrated jet grouting columns. Jet-grouted columns

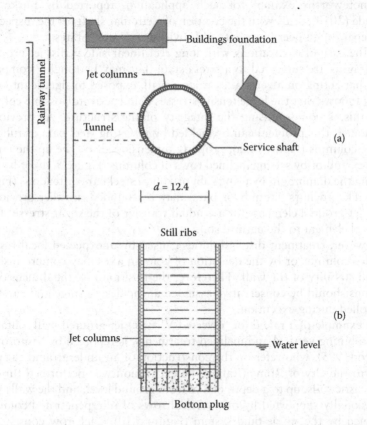

Figure 5.8 Example of provisional support of shaft excavation by means of jet grouting: (a) plan view; (b) cross-section. (From Balossi Restelli, A. et al., Tecnologie speciali per il preconsolidamento di scavi nelle alluvioni di Milano in occasione della costruzione della linea 3 della Metropolitana, In Proceedings of the International Conference 'Grandi Opere Sotterranee', Florence, Italy, June 8–11, 1986: pp. 612–618 [in Italian], 1986.)

having nominal diameter of 60 cm were placed at a centre-to-centre distance of 50 cm. After completion of jet grouting treatments, the shaft was excavated, and the cylindrical supporting element was finished by a layer of steel mesh and shotcrete, having 20-cm thickness, and was supported with steel ribs placed at 1-m intervals. In the last few metres, the excavation was performed below the groundwater level, after having sealed the bottom of the shaft by means of conventional low-pressure grouting. However, the deviations of the drilling axes with respect to the vertical direction resulted in the imperfect penetration of the columns, at great depth, and it was thus necessary to clog the spaces between the jet columns with further low-pressure grouting. The project was, anyway, successful in containing the building settlements below the value of 1.5 mm.

A noteworthy example of such application, reported by Burke and Yoshida (2013), deals with the creation of a circular shaft for the inspection and repair of a sewer in the city of Providence (United States).

In the case of excavations with long rectilinear side walls, jet grouting diaphragms are subjected to a stress state that may be roughly compared with that acting on a vertical cantilever wall, exposed to significant bending. It follows that the large tensile stresses could activate possible collapse mechanisms compromising the integrity of the structure. As previously mentioned, this problem can be tackled by inserting steel bars or tubes in the jet columns (Figure 5.9a), properly dimensioned to take up the tensile stresses, and/or by setting inclined rows of columns (Figure 5.9b) or by supporting the diaphragm by using subhorizontal steel bars or tendons (Figure 5.9c). The anchors (strands or bars) may be founded in other previously made jet-grouted elements for a gradual transfer of the shear stresses from the steel element to the natural soil.

However, treatment discontinuities caused by unexpected local restrictions of columns or by the deviation of column axes may compromise the overall stability of the wall. Therefore, the position and the dimension of columns should be conservative assigned at the design stage and carefully controlled during execution.

An example of a raked (or 'inverted V' type) jet-grouted wall, obtained by combining vertical and inclined treatments, is reported by Santoro and Bianco (1995), who refer on the construction of an underground car park close to the city of Rome (Italy). The excavation was performed through pyroclastic soils, up to a depth of 8 m from ground level, and the walls were provisionally supported by two vertical rows of interpenetrated columns, obtained by the single-fluid system (Figure 5.10). Each row consisted of 16-m-long columns of nominal diameter equal to 0.70 m, reinforced with a steel tube and located at a 0.60-m centre-to-centre distance, giving rise to a vertical supporting element with a thickness of approximately 1.90 m. The element was linked to an additional row of steel-reinforced columns,

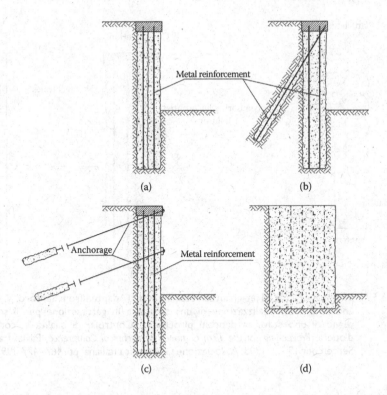

Figure 5.9 Jet-grouted retaining structures: (a) vertical cantilever; (b) raked or 'inverted V' type; (c) anchored; (d) massive.

having the same length, inclined at 25° to vertical. Finally, all columns were connected on top by a longitudinal beam of reinforced concrete.

A typical case of anchored jet-grouted structure is presented by Sondermann and Toth (2001), who refer on an excavation made in the vicinity of Bonn (Germany), within a bank of alluvial soils composed of two layers, respectively, of sand and sandy gravel. The excavation was driven to a depth of 13.30 m from ground level, performed adjacent to some four- and five-storey buildings (Figure 5.11a). The supporting jet grouting element was obtained by creating slightly inclined columns and was then provided with four rows of anchors to reduce bending. The treatments were performed with the triple-fluid system, producing columns with mean diameters of approximately 1.2 m. The groundwater level was 5.2 m higher than the bottom of the excavation. Therefore, the jet grouting treatments were protracted up to a depth of 18 m from ground level to reach a layer of very consistent clays, thus avoiding water flow into the excavation. A similar application, reported by Garassino (1983), presents an excavation of

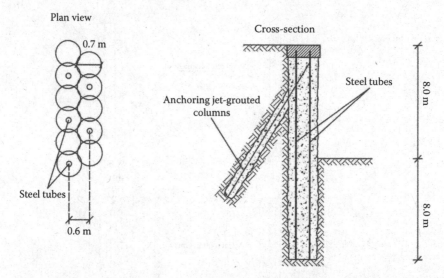

Plan view

0.7 m

Cross-section

Steel tubes

8.0 m

Anchoring jet-grouted
columns

Steel tubes

0.6 m

8.0 m

Figure 5.10 Example of a raked jet-grouted retaining wall. (Adapted from Santoro, V. M. and B. Bianco, Realizzazione di una struttura in gettiniezione per il sostegno di uno scavo in depositi piroclastici: Controllo di qualità in corso d'opera. *Proceedings of the 19th National Geotechnical Conference*, Pavia, Italy, September 19–21, 1995, Associazione Geotecnica Italiana: pp. 467–477, 1995.)

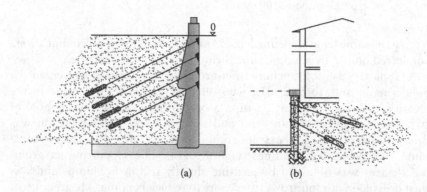

(a)   (b)

Figure 5.11 Examples of anchored jet-grouted retaining structure. ((a) Adapted from Sondermann, W. and P. S. Tóth, State of the art of the jet grouting shown on different applications. *Proceedings of the 4th International Conference on Ground Improvement*, Helsinki, Finland, June 7–9, 2001, Finnish Geotechnical Society: pp. 181–194, 2001; (b) adapted from Garassino, A. L., Uso delle colonne in terra stabilizzata con jet grouting come elemento provvisionale. *Proceedings of the 15th National Geotechnical Conference 2*, Spoleto, Italy, Associazione Geotecnica Italiana: pp. 155–161 [in Italian], 1983.)

up to approximately 8.50 m below ground level, supported by two parallel rows of jet-grouted columns having 0.70-m diameter, partly overlapped and reinforced with an inner tube of 8-cm diameter and 4-mm thickness.

The earth-retaining structure was, in this case, supported by a double order of jet-grouted anchors and a concrete basement slab aimed at reducing the bending moments in the jet-grouted columns.

5.4 WATER BARRIERS

Jet grouting is often adopted to provide permanent or provisional water flow barriers in the subsoil (Burke 2007a; Burke and Yoshida 2013). In fact, permanent waterproofing is frequently required to avoid or restrict seepage under hydraulic structures, such as dams and weirs, or to ensure sealing of contaminated sites and landfills. Provisional waterproofing is also required to avoid water flows into excavations to be performed through pervious soils under the water level. In all cases, effective water barriers can be obtained by creating jet-grouted diaphragms (cutoffs) and/or horizontal jet-grouted slabs (bottom plugs).

With regard to the layout of vertical jet-grouted diaphragms, several solutions have been used or proposed (Bell 1993). A collection of design schemes is given in Figure 5.12, where cutoff types a, b and c are obtained

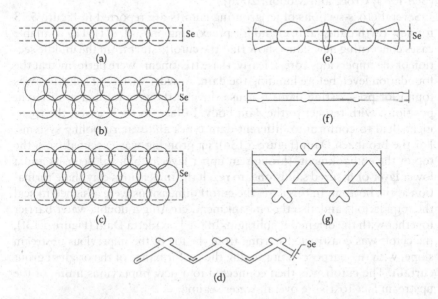

Figure 5.12 Typical arrangements of waterproofing diaphragms made of jet columns and panels.

by joining one or more parallel rows of cylindrical, partially overlapped columns, whereas type d is achieved by connecting V-shaped jet-grouted panels, each of them created by performing the treatment without monitor rotation. The panels can be combined according to various possible shapes (linear, wedge, cellular), as shown in Figure 5.12 (f, g and h). Overlapped candy-shape elements (type e) can be obtained by partly rotating the monitor during injection (see Chapter 2).

The choice of the most appropriate diaphragm layout is clearly guided by the degree of confidence placed by the designer on the effectiveness of the treatment procedure, as related to the specific project requirements. It is noted, however, that jet-grouted panels are used only in special cases, whereas the most popular solutions are obtained by using cylindrical columns and by choosing the distance between treatments as low as needed for ensuring the continuity of the diaphragm. Particular care must be paid in controlling the construction sequence (fresh-in-fresh or primary-secondary sequence) according to the indications provided in Chapter 2.

Jet grouting cutoffs have been used for several types of barrages, such as earth dams and cofferdams, concrete gravity dams and concrete diversion weirs (Croce and Modoni 2005, 2008). In some cases, the cutoffs were foreseen at the design stage and carried out before the construction of the dam (Attewill et al. 1992; Sembenelli and Sembenelli 1999; Guatteri et al. 2012); in other cases, the treatments were accomplished on existing dams, with the aim of improving the efficiency of existing waterproofing barriers (Bell 1993; Croce and Modoni 2006).

Several cross-sections of jet grouting cutoffs are reported in Figure 5.13 for new earth dams (a, b, c) and for preexisting ones (d, e, f). The available cases concerning new dams show that the cutoffs start from the middle section of the impervious core. Clearly, these treatments were performed at the foundation level, before building the dam. A wider variety of solutions was found for pre-existing dams because the cutoffs were located at different positions, with respect to the dam body. In fact, in these cases, the treatments had to conform to different dam types and waterproofing systems. For the Bronbach Dam (Figure 5.13d), jet grouting was executed from the top of the embankment, through an upper layer of alluvial deposits and a lower layer of fissured sandstone, to reach the impervious clay shale formation at the bottom. In this case, the cutoff dimensions were assigned to seal the foundations and also the embankment, creating a double water barrier together with the original diaphragm. In the Forcoletta Dam (Figure 5.13f), the cutoff was constructed starting from the toe of the impervious upstream slope, with the purpose of improving the performance of the original grout curtain. The cutoff was then connected to a new impervious lining of the upstream face to assure overall water sealing.

From the available sample of cases, it appears that jet grouting cutoffs are most frequently used as a mean for fixing existing dams. This trend

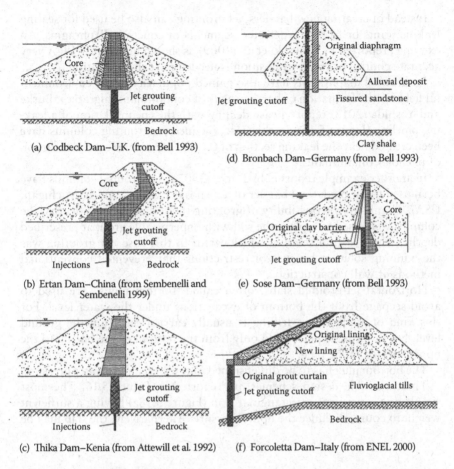

(a) Codbeck Dam–U.K. (from Bell 1993)

(d) Bronbach Dam–Germany (from Bell 1993)

(b) Ertan Dam–China (from Sembenelli and Sembenelli 1999)

(e) Sose Dam–Germany (from Bell 1993)

(c) Thika Dam–Kenia (from Attewill et al. 1992)

(f) Forcoletta Dam–Italy (from ENEL 2000)

Figure 5.13 Cross sections of jet grouting cutoffs for new dams and for pre-existing earth dams. (Adapted from Croce, P. and G. Modoni. 2006. *Ground Improvement* 10(1): pp. 1–9.)

is probably caused by the inherent flexibility of jet grouting. In fact, by assembling the jet columns with proper arrangements, the cutoffs can be customised to different dam types and geological settings. Another relevant consideration is that jet grouting treatments can be performed under unfavourable logistic conditions because heavy equipment is not required.

Defects in jet-grouted diaphragms can occur as a result of

- Insufficient overlapping of individual elements
- Jet shadows caused by natural or man-made obstacles or by previously created columns
- Inhomogeneous ground conditions

Instead of creating new barriers, jet grouting can also be used for sealing leaking joints or to repair defected segments of concrete diaphragms. An example, reported by Ryjevski et al. (2009), is shown in Figure 5.14. A very accurate control of columns position is needed in such application.

Jet-grouted diaphragms have also gained popularity in the environmental field for the formation of barriers against contaminant migration. Burke and Yoshida (2013) report a case dealing with the encapsulation of a leaking pipeline (Figure 5.15). In this case, inclined jet grouting columns have been created near the leaking section of the pipeline to prevent the diffusion of phenole contaminant.

In another example reported by Burke (2007a), jet-grouted columns have been used to form a lateral barrier of a disposal area at Dundee (Michigan, USA). Thanks to the possibility of grouting the soil at selected depths, the columns were used to replace the soil with impermeable grout at prescribed depths, where a sand stratum was located. In this case, jet grouting was the winning solution because of restrictions that prevented conventional methods of wall construction.

Horizontal jet-grouted slabs, often called 'bottom plugs', are used to avoid seepage from the bottom of excavations under the water level. For this kind of application, drilling is usually carried out from the ground level, but grouting is performed only from the bottom of the boring to the upper surface of the plug.

The bottom plug has to be dimensioned to resist the uplift water pressure.

Three possible design schemes are reported in Figure 5.16. The most simple one consists of a rectangular slab (Figure 5.16a) having a sufficient weight to counterbalance the uplift pressure. Alternatively, the slab may be

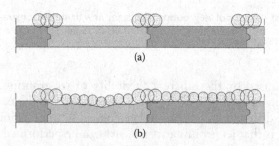

Figure 5.14 Example of jet grouting solution used for sealing defective concrete diaphragms: (a) repair of joints; (b) repair of defected segments. (Adapted from Ryjevski, M. et al., Jet grouting application for Dubai Metro Construction. *Proceedings of the World Tunnel Congress*, Budapest, Hungary, May 23–28, 2009, Paper O-10-19: 10 p., 2009.)

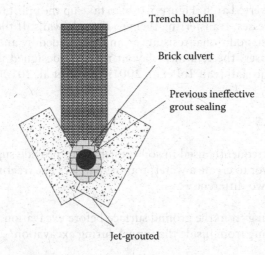

Figure 5.15 Example of a jet grouting used for the formation of barriers against contaminant diffusion. (Adapted from Burke, G. K. and H. Yoshida, Jet grouting, In Kirsch, K. and Bell, A. eds., *Ground Improvement*, 3rd ed.: CRC Press Taylor & Francis Group: pp. 207–258, 2013.)

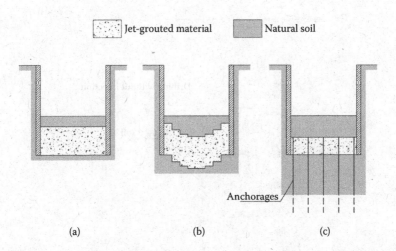

Figure 5.16 Bottom-plug schemes made by jet grouting: (a) plain slab, (b) inverted arch and (c) anchored slab.

shaped as an inverted arch (Figure 5.16b) to take up the uplift force through compressive stresses, transferring them to the side walls. If the uplift pressure is very high, additional resistance can be provided by anchors (Figure 5.16c). In all cases, the treatment layout must be designed to assure the continuity of the slab (van Tol et al. 2001; Eramo et al. 2012).

5.5 TUNNELS

Jet grouting is frequently used in soil tunnelling to provide support during excavation and/or to create a waterproofing barrier. The treatments may be carried out in two different ways:

1. By operating from the ground surface before excavation
2. By operating from inside the tunnel during excavation

The first construction method (Figure 5.17) has the advantage of separating soil treatment from excavation, thus reducing the overall construction time by proper planning (Arroyo et al. 2012). Although this scheme has the advantage of allowing jet grouting independently from the tunnel construction sequence, treatment can be conveniently performed from the ground level only if the soil cover is thin and if the ground surface is easily accessible. Therefore, in most cases, it is necessary to perform the treatments by working inside the tunnel.

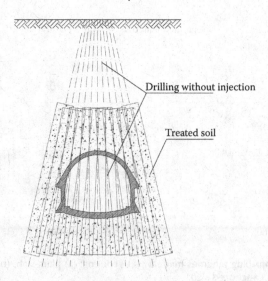

Figure 5.17 Tunnel support by jet grouting performed from the ground surface.

The use of jet grouting operating from inside the tunnel is related to the so-called 'canopy technique' (Figure 5.18). This technique is based on the use of various complementary soil reinforcement methods, which are set with particular geometrical and sequential features to provide the stability of the tunnel contour and face during excavation. Such a canopy can be obtained by means of jet-grouted columns and/or steel micropiles that are installed in advance with respect to excavation. For very weak soils, the stability of the tunnel face may be granted by diffused reinforcement of the soil behind the face, which can be pursued either by jet grouting or by of fibreglass elements (bars or tubes). In fact, face treatment is rather expensive and time consuming, but it can provide effective support for unstable excavation and minimise surface settlements as well. During excavation, the face-reinforcing components are progressively removed together with the soil, where the contour elements are contemporaneously supported by setting up steel ribs and shotcrete to complete the provisional lining of the tunnel. The final lining, made of reinforced concrete, may then be installed at a later and more convenient time.

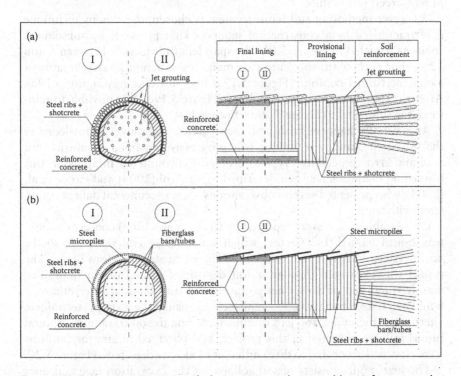

Figure 5.18 Schematic drawings of the canopy technique: (a) reinforcement by jet-grouted columns; (b) reinforcement by steel micropiles and fibreglass bars or tubes.

In tunnelling applications, jet grouting is usually performed by the single-fluid method because it does not require heavy equipment and is thus more dependable for underground operations. The adoption of single-fluid jet grouting is also due to security reasons because the single-fluid method does not require the use of compressed air jets, which might induce serious risks to the workmanship in narrow underground spaces.

In principle, each reinforcement technique is best suited for particular soil types and seepage conditions. Jet grouting treatments are generally preferred for sandy soils, providing larger columns that can be set aside, forming a sort of supporting arch (Figure 5.18a). The arch can also provide waterproofing, when properly dimensioned, if pore-water pressures are not too high. For fine-grained materials, single-fluid jet grouting is not very effective, and the use of micropiles is generally preferred (Figure 5.18b). In particular, steel micropiles are used for the tunnel contour, whereas fibre-glass bars or tubes are used for face reinforcement because they can easily be truncated during excavation. In difficult cases or doubtful soil conditions, it may be useful to combine jet grouting and micropiles to form a sort of reinforced jet column.

Whatever method of soil reinforcement is chosen, the canopy technique is characterised by a construction sequence that proceeds by subsequent spans as depicted in Figure 5.19. The span length is usually between 6 and 12 m. For each span, there are two main construction phases: treatment along the tunnel contour (Figure 5.19a) and soil digging (Figure 5.19b). After the span excavation is completed (Figure 5.19c), it is possible to reinforce the tunnel face (Figure 5.19d) if needed.

As it usually happens with new construction techniques, knowledge in the use of the canopy technique was progressively gained by means of a trial-and-error procedure. Typical problems encountered in applying this tunnelling method have been described by van Tol (2004) and Croce et al. (2004), who presented several observations and experimental data accumulated with time.

One notable case history reported by Croce et al. (2004) concerns a highway tunnel named 'Les Cretes', which was built several years ago in the northwestern Italian Alps, between the city of Aosta and Mont Blanc. The tunnel was excavated through a soil deposit of glacial origin (moraine), mainly composed of dense sandy gravels and silty sands with erratic rock boulders. Excavation was accomplished by using the canopy technique, alternating micropiles and jet grouting according to soil conditions. Several tunnel failures occurred in this project, and observed failure mechanisms were classified according to three different modes, as depicted in Figure 5.20.

The first mode consists of soil collapse at the excavation face and does not involve the canopy or the provisional lining (Figure 5.20a). This failure mode can be attributed to weak layers of soil and/or piping induced by water seepage. The most logical countermeasure for preventing such

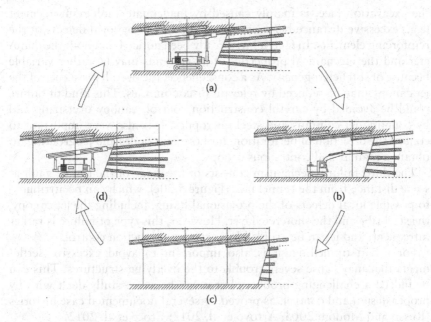

Figure 5.19 Tunnelling sequence by the canopy technique: (a) reinforcement of the tunnel contour by jet grouting and/or steel micropiles; (b) excavation; (c) excavation completed; (d) face reinforcement by jet grouting and/or fibreglass bars or tubes (optional).

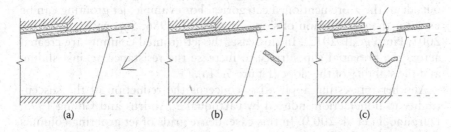

Figure 5.20 Typical failure mechanisms of jet-grouted canopies. (a) Face failure; (b) tip failure; (c) lining failure.

a failure mechanism consists of face reinforcement capable of providing tensile strength to the soil mass behind the excavation face. However, a suitable stabilising effect could also be obtained by increasing the distance from the tunnel face to the canopy tip, that is, by making longer columns or by reducing the excavation span.

The second failure mechanism consists of the collapse of the canopy tip, just behind the tunnel face (Figure 5.20b). Such failure, which also involves the excavation face, is mainly caused by inadequate vault reinforcement (e.g., excessive distance between the micropiles) or by local defects of the reinforcing elements. In fact, it has already been noticed that both the diameter and the mechanical properties of jet columns may be rather variable because of soil heterogeneity. As a consequence, the overall resistance of the jet canopy may be reduced by relevant discontinuities. This kind of failure could be avoided by careful construction control, canopy oversizing and by coupling jet columns and steel micropiles. It is also very important to control the direction of perforation, for both micropiles and jet grouting, to obtain a regular and continuous canopy.

The third failure mechanism consists of breakage of the tunnel vault at some distance from the tunnel face (Figure 5.20c), which can be attributed to possible local defects of the provisional lining, including the jet canopy, the steel ribs and the shotcrete layer. However, this type of failure is rather infrequent, and it can be avoided by careful construction control.

For urban tunnelling, it is also important to avoid excessive settlements that may cause severe trouble to the overlying structures. This can be indeed a challenging problem, but it can be successfully dealt with by proper design and control, as proved by several documented case histories (Russo and Modoni 2005; Arroyo et al. 2012; Croce et al. 2012).

5.6 OTHER APPLICATIONS

There are other less frequent examples of jet grouting applications falling outside of the aforementioned categories. For example, jet grouting can be used for the stabilisation of slopes (Langbehn 1986; Sopeña Mañas et al. 2001; Pinto et al. 2012). In this case, the jet-grouted columns are created across the potential slip surface to increase the resistance against sliding and the stability of the slope (Figure 5.21a).

Another interesting application concerns the reduction of the susceptibility to liquefaction induced by earthquake (Andrus and Chung 1995; Durgunoğlu et al. 2003). In this case, dense grids of jet grouting columns are created at depths where soil is susceptible to liquefaction to minimise soil deformation and the consequent development of pore pressure induced by earthquakes. It is worth noting that such phenomena occur typically in loose sandy soils, which are particularly suitable to jet grouting treatment. An example of such application is shown by Burke and Yoshida (2013), dealing with the seismic remediation of the Wickiup Dam (Oregon, USA). In this case, the jet-grouted columns were created on the left abutment of the dam, which was resting on two separate layers of diatomaceous silt and volcanic ash likely to liquefy (Figure 5.21b). Jet grouting, in this case, allowed *in situ* stabilisation during normal reservoir operations,

(a)

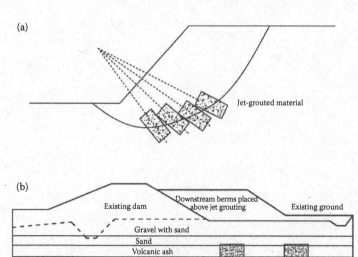

Jet-grouted material

(b)

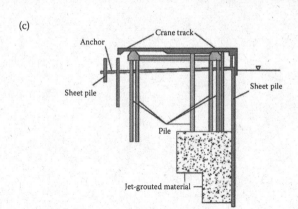

Existing dam

Downstream berms placed
above jet grouting

Existing ground

Gravel with sand

Sand

Volcanic ash

Diatomaceus silt

Dense silt and sand

Diatomaceus silt

Dense silt and sand

Jet-
grouted
material

(c)

Crane track

Anchor

Sheet pile

Sheet pile

Pile

Jet-grouted material

Figure 5.21 Jet grouting applied to the stabilisation of slopes (a), the remediation of liquefaction (b) and the protection of waterfront structures (c).

thus reducing the inherent risk associated with an excavate-and-replace alternative.

In recent times, jet grouting has also gained popularity in the field of marine structures. In the example reported in Figure 5.21c, a massive jet grouting treatment has been performed to reinforce the footing of a wharf and to allow a further excavation of the sea bottom.

Chapter 6

Design principles

6.1 BASIC CONSIDERATIONS

6.1.1 Design goals

As already pointed out in Chapter 5, jet grouting is frequently used for many different goals by arranging the columns in various possible ways to create one-, two- or three-dimensional elements. It follows that present design procedures for foundations, retaining structures, water barriers, tunnels, etc., should be properly adapted to encompass the use of jet grouting as a possible alternative to conventional bored or driven rigid elements (e.g., piles, walls, etc.). Table 6.1 summarises the most frequent applications of jet grouting, pointing out the role played by the jet-grouted structure, the typical geometrical configurations, the needed geometrical and mechanical characteristics of the columns and the assessments to be carried out.

It seems logical that the design of jet-grouted structures, like for any other structure, should be developed following a number of logical steps, going from site characterisation to cost assessment and passing through the verification of serviceability and ultimate limit states. However, with respect to conventional geotechnical structures, it is commonly recognised that the technological aspects of jet grouting play a more relevant role. It follows that, in common practice, the designer provides only simple indications on the jet grouting project, which is then specified only at the construction stage, by following some sort of trial-and-error procedure on site. The reason for such simplistic procedure lies in the widespread belief that the effects of jet grouting cannot be forecasted at the design stage, and so, it may seem more realistic to rely only on a purely empirical approach. However, even if this lack of knowledge was probably true in the pioneering age of jet grouting, the experience gained all over the world in more than 30 years of practice and the recent research activity on this topic have now provided more reliable design tools. Clearly, there are additional

Table 6.1 Requirements and assessment of typical jet grouting applications

Applications	Functions	Possible layouts	Geometrical requirements	Properties of the jet-grouted material	Design calculations
Foundations	-Bearing capacity increment -Settlements reduction	-Grid of isolated columns -Massive body formed by overlapped columns	-Minimum column diameter -Span/diameter of columns	-Strength/stiffness -Strength/stiffness	-Bearing capacity -Settlements -Resistance of the jet-grouted material
Earth-retaining structures	-Earth support -Earth pressure reduction	-Single or multiple rows of overlapped columns -Massive body formed with overlapped columns	-Continuity -Minimum overall thickness -Continuity -Span/diameter of columns	-Strength/stiffness -Strength/stiffness	-Stability -Displacements (when required) -Continuity -Resistance of the jet-grouted material -Stability -Displacements (when required) -Resistance of the jet-grouted material
Tunnels	-Provisional support on tunnel contour and/or face -Provisional waterproofing	-Frustum of cone-shape canopies -Massive body made of closely spaced or overlapped columns -Reinforcement from the ground level	-Continuity -Minimum thickness -Continuity -Span/diameter of columns -Continuity -Minimum thickness	-Low permeability -Strength/stiffness -Low permeability -Strength -Low permeability -Strength/stiffness	-Stability -Displacements (when required) -Continuity -Resistance of the jet-grouted material

Cutoffs	-Seepage barrier	-Single or multiple rows of overlapped columns	-Continuity	-Low permeability	-Continuity
Bottom plugs	-Provisional waterproofing -Uplift resistance -Base restraint of diaphragm walls	-Massive body made of overlapped columns	-Continuity	-Low permeability -Strength	-Seepage -Resistance against heave -Continuity -Resistance of the jet-grouted material
Improvement of existing structures	-Underpinning	-Isolated columns -Massive body made of closely spaced or overlapped columns	-Minimum diameter of columns -Span/diameter of columns	-Strength/stiffness	-Stability -Displacements/settlements -Resistance of the jet-grouted material
	-Earth pressure reduction	-Massive body made of closely spaced or overlapped columns	-Span/diameter of columns	-Strength/stiffness	-Stability -Resistance of the jet-grouted material
	-Waterproofing	-Overlapped columns	-Continuity	-Low permeability	-Seepage -Continuity
	-Integration/ repair of previous treatments	-Various shapes	-Span/diameter of columns	-Strength/stiffness -Low permeability	-Variable

steps that must be added to the usual design process, strictly related to the quantification of the technological effects:

- The choice of the jet grouting procedure (see Chapter 2)
- The quantification of treatment parameters (see Chapters 2 and 3)
- The prediction of the dimensions and the mechanical properties of the jet grouted columns (see Chapter 4)
- The analyses of possible undesired collateral effects on the surrounding constructions and on the environment (see Chapter 5)

Because technology plays a relevant role in the success of a jet grouting project, the designer should be aware of all the previously listed steps, not blindly leaving them to the specialist contractor only. The prediction of the performance of a jet-grouted structure with regard to ultimate and serviceability limit states must thus be seen as a whole, with the definition of the set of operations necessary to obtain the desired result in terms of column dimensions and properties. In this sense, a winning design strategy must conjugate theoretical and technological knowledge to conceive practically feasible solutions able to provide safety, functionality and economy.

Suggested design procedures and possible alternative approaches, methods of analysis and calculation examples are reported in this chapter for the most common applications of jet grouting, showing that the effects of technology on structural performance can be rationally considered and accounted for.

6.1.2 Guidelines and codes of practice

Indications given by the existing rules, codes of practice and guidelines on ground improvement techniques and, particularly, on jet grouting, are not homogenous around the world.

The European Committee for Standardisation (Committé Européen de Normalisation [CEN]) has set forth in time a number of European Standards (or Euro Norms, often referred to as 'ENs'). Among these European Standards, CEN has issued some standard rules for the 'Execution of Special Geotechnical Works'. Indications on jet grouting are given by the Execution Standard EN 12716 (*Execution of special geotechnical works—Jet grouting*, 2001), which describes a sequential list of activities that should be followed in a typical jet grouting project, as reported in Table 6.2.

EN 12716 states that not only the properties (geometric, mechanical and/or hydraulic) of the jet-grouted soil elements have to be quantified, but also the procedure to obtain and measure these properties should be specified. When, because of a lack of experience or a limited amount of investigations, these properties cannot be quantified up to a satisfactory degree of

Table 6.2 Recommended list of activities in the design and execution of jet grouting (EN 12716)

Number	Activity
1	Provision of site investigation data for the execution of jet grouting works
2	Decision to use jet grouting, preliminary trials and testing if required; provision of a specification
3	Acquisition of all legal authorisation necessary for the execution from authorities and third parties
4	Overall design of jet-grouted structure and definition of the geotechnical category[a]
5	Consideration of the relevant temporary phases of execution
6	Assessment of the site investigation data with respect to design assumptions
7	Assessment of the construction feasibility of the design
8	Execution of trials if required and of any relevant tests
9	Evaluation of the results of the preliminary trials and tests
10	Selection of a jet grouting system
11	Assessment of the jet grouting system and definition of the working procedures
12	Definition of the dimensions, location and orientation of jet-grouted elements
13	Instructions regarding the working sequence if required
14	Definition of the working sequence
15	Instruction to all parties involved of key items in the design criteria to which special attention should be directed
16	Specification for monitoring the effects of jet grouting on adjacent structures (type and accuracy of instruments, frequency of measurement) and for interpreting the results
17	Definition of tolerable limits of the effects of jet grouting works on adjacent structures
18	Execution of jet grouting works, including monitoring of the jet grouting parameters
19	Supervision of the works, including the definition of the quality requirements
20	Monitoring the effects of jet grouting works on adjacent structures and presenting the results
21	Control of the quality of works

[a] Geotechnical categories are defined in Eurocode 7 (EN 1997-1 [2004]).

accuracy, demonstrative field tests shall be executed to prove the efficiency of treatments and to refine the treatment procedure.

However, being devoted to the execution rather than the design procedure, EN 12716 provides only some general recommendations on design, which are briefly exposed in a section entitled 'Design considerations'. This section points out the need of careful investigations of the site conditions to assess the applicability of jet grouting.

EN 12716 also states that, given the typical variability of jet-grouted elements, statistical analyses could be useful to estimate the design values of jet-grouted elements. Therefore, although it is not aimed to deal with design, EN 12716 recognises the relevance of the technological peculiarities of jet grouting for design calculations.

The topic of geotechnical design is tackled, for the European countries, by the Eurocode 7 (*EC7—Geotechnical Design*), which is the reference document for the member states of the European Union. However, in its present version, EC7 provides only some generic indications on the design of ground improvement and ground reinforcement projects. It is thus expected that more specific design procedures will be provided in the next version of EC7.

In the meantime, some European nations have set forth their national specific guidelines on jet grouting. In particular, Germany has issued specific rules on the design of ground improvement by means of jet grouting, deep mixing and grouting (DIN 4093:2012-08). Specific technical recommendations on jet grouting have been developed in Italy (AGI 2012). These recommendations suggest that the design path should be organised in an ordered sequence of experimental activities, data processing and calculations, and also provide some quantitative indications on the most important design parameters.

In Japan, national guidelines have also been issued (Japanese Guidelines on Jet Grouting; JJGA 2005). Such guidelines encompass indications on design, proposing a strongly deterministic approach. The procedure suggested in these guidelines is articulated in the following steps:

1. A set of parameters must be initially determined for the characterisation of soil, this latter classified as ordinary (clayey and sandy) and special (sandy gravel and organic). Among all parameters, a predominant role on the definition of the treatment procedure is played by the number of blows obtained by Standard Penetration Tests (N_{SPT}).
2. Suitability of the two most used Japanese construction methods (double-fluid jumbo special grout (JSG) and triple-fluid jet grouting) is estimated based on the soil type and the N_{SPT} value.
3. Values of the diameter of influence (i.e., column diameter) are provided for the different methods (JSG and jet grouting method) and soils (coarse and fine), depending on the N_{SPT} value. Specific values of the lifting speed of the injection rod and of the grout flow rate are given to obtain the diameter of influence. In case of soil heterogeneity, the most conservative soil condition must be assumed.
4. Design values of the mechanical properties (unconfined compression strength, cohesion, bending/tensile strength and Young modulus) of

the jet-grouted material are given based on the combination between the soil type and the jet grouting technology.

5. Layouts (plan arrangement and spacing of columns, thickness of structures) are given for the most common applications (cutoff, bearing capacity improvement, starting and arrival sections of shield tunnels, filling gaps between earth-retaining structures).

6. Safety factors are given for each application, distinguishing if the jet-grouted structure has a temporary or a permanent function (Table 6.3).

7. Composition and dosage of materials (water, cement, mineral and chemical additives) are given for the different applications. With regard to the additives products from a list of certified companies should be used.

Table 6.3 reports the safety factors proposed by the Japanese Guidelines. With this approach, the design and execution of jet grouting are framed into a very constrained scheme, and even the geometrical choices (grid array and spacing) are taken out of the responsibility of the designer.

A totally different approach is followed in the guidelines for jet grouting prepared by the American Society of Civil Engineers (GI-ASCE 2009), which are more generic and essentially devoted to identifying the role and qualification of the different operators and to rule the relationship between the client and the contractor rather than defining a design procedure. With regard to the design tasks, the ASCE guidelines specify that

Table 6.3 Safety factors for different jet-grouted structures

Application	Objectives of improvement	Factor of safety	Note
Improvement at the bottom of open-cut excavation	Heaving protection	1.5	Factor of safety for the permanent structure should be equal to three or more
	Boiling protection	1.5	
	Designing the penetration depth	1.5	
Starting section of shield tunnelling	Protection of cutting face or reaction wall	1.5	
Arrival section of shield tunnelling	Cutting face protection	1.5	
	Tail section protection	1.0	
Soil protection at the gap between earth-retaining walls	Combined with soldier beam	1.0	
	Jet grouting only	2.0	
Caisson-type pile	Reinforcement of sidewall	1.5	
	Cutting face protection	2.0	

Source: Modified from JJGA, *Jet Grouting Technology: JSG Method, Column Jet Grouting Method.* Technical Information of the Japanese Jet Grouting Association, 13th ed. (English translation), October 2005: 80 p.

'the contractor shall be responsible for selection of jet grouting parameters, equipment, and construction of the jet-grouted elements to meet the design intent', whereas 'the engineer is responsible for the overall design of the jet-grouted soil element or soil cement structure'. Therefore, these guidelines clearly separate the definition of a set of operative parameters to perform jet grouting treatment from the assessment of the performance of the jet-grouted structure, assigning them respectively to the contractor and the client. Particular attention is devoted to the quality control and quality assurance procedure, and to the list of operations that should be performed by the contractor and certified by the client (e.g., type and frequency of tests, acceptance criteria).

In this chapter, the design philosophy proposed by the Italian Recommendations on Jet Grouting (AGI 2012) is encompassed, and reference will be made to the logical pattern summarised by the flowchart reported in Figure 6.1.

The first design step reported in the flowchart consists of performing geotechnical investigations aimed at the characterisation of the subsoil. Particular attention must be paid to the determination of the properties that play a relevant role in the jet grouting process and to the examination of the stratigraphic conditions so as to identify the variations, even local, of the soil characteristics able to affect the properties of jet-grouted elements. These experimental data should be carefully examined on the basis of previous experiences to determine whether a given soil is suitable for jet-grouted treatment. Heavily overconsolidated clays, cemented soils and rocks are typical examples in which the usefulness of jet grouting is questionable if not nil. Thereafter, logistic restrictions and environmental compatibility should be evaluated with reference to the possible operative conditions (treatment from the ground surface or performed underground, etc.).

A limited availability of space for the access and movement of machineries in the working area, the nearby presence of sensitive buildings or structures and strong environmental constraints limiting the management of the spoil may significantly influence the choices or may even discourage the use of jet grouting.

In case all previous investigations give positive answers, a preliminary choice of the treatment system (single, double or triple fluid) can be made. Having thus defined the soil properties and the treatment parameters, the correlations introduced in Chapter 4 may be used to predict the mean diameters and the mechanical properties of the columns. At this stage of design, different possible diameters can be considered, and tentative layouts can be sketched by fixing the number of elements and their geometrical arrangement. Each of these solutions should then be checked by verifying with proper calculations the compliance of the jet-grouted structure to ultimate and serviceability limit states. Obviously, the ideal solution should optimise the combination of safety, functionality, economy and feasibility.

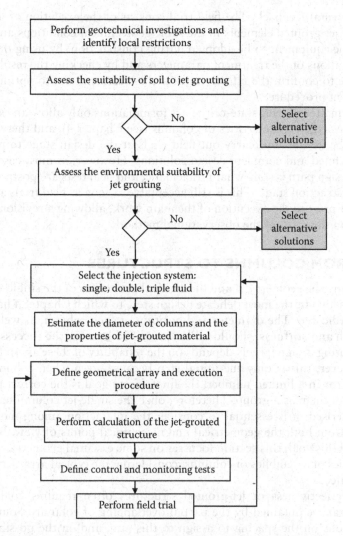

Figure 6.1 Sequence of design activities suggested for a jet grouting project.

The design report should also include the planning of the control and monitoring activities to be carried out during construction and defining the experimental methods, acceptance criteria and thresholds to be respected to certify the success of the jet grouting project.

In the early stages of construction, before starting the jet grouting production, the correctness of the assumptions made in the design should be verified. To this aim, a field trial (see Chapter 8.1) should be executed, unless experience previously gained in the same site give the needed experimental feedback. The field trial consists of the execution of demonstrative jet-grouted elements in the same geotechnical conditions and with the same equipment to be adopted during construction. By using different combinations of the treatment parameters and by checking the results, it is possible to confirm the effectiveness of the treatment and to optimise the treatment procedure.

As a matter of fact, state-of-the-art formulations only allow an estimate of the average characteristics of columns (see Chapter 4), and therefore, it would be desirable to carry out field trials at the design stage to provide more refined and complete design solutions. However, in most cases, this ideal design path is somewhat problematic, and field tests are postponed to the construction stage. This is still acceptable as long as field trials are carried out prior to the execution of the main work, allowing a revision of the design in the spirit of the observational method.

6.2 FROM COLUMNS TO STRUCTURES

Assigning the geometrical and mechanical properties of the single-jet columns is, by far, the most delicate design step, to which Chapter 4 has thus been dedicated. The diameter and position of each column, as well as its strength and stiffness, should all be quantified because the success of the jet grouting design largely depends on the reliability of these assumptions.

However, this is only the initial step because, even when columns are isolated or in a limited number, the final design goal is the correct performance of them as a group. Therefore, once the single-jet column has been characterised, it is essential to consider the interaction among more columns, from both the geometrical and mechanical points of view. Bearing in mind this goal, this section focusses on some essential geometrical characteristics of assemblies of columns, considering the typical arrays adopted in practice.

The effectiveness of jet-grouted structures often requires continuity, which can be obtained by the partial overlapping of columns. Some considerations on the spacing to assign to this aim, and on the possible columns grid array, can be thus useful to reach the best possible compromise between cost containment (spacing as high as possible) and continuity (spacing not too high).

To start with, let us introduce some trivial but necessary geometrical considerations on partially overlapped columns. In principle, the overlapping of two identical, adjacent and parallel columns can be ensured if their axes are positioned at a distance (spacing) smaller than their diameter (Figure 6.2a). For cylindrical columns, the thickness of the element formed by the two columns is not constant. From now on, the minimum thickness t (see Figure 6.2a) will be considered as the reference value. For the simplest case

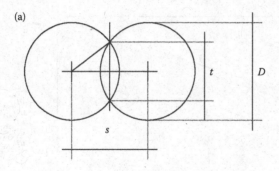

(a)

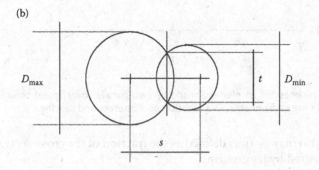

(b)

Figure 6.2 Definition of spacing *s* and thickness *t* for an element made by two parallel overlapped columns of equal (a) and different (b) diameters.

of two columns having the same diameter D and placed with a spacing s, the thickness t can be calculated as

$$t = \sqrt{D^2 - s^2} \tag{6.1}$$

whereas, in the case of two columns having different diameters ($D_{min} < D_{max}$; Figure 6.2b), the relationship between the spacing, diameters and thickness can be written as

$$\frac{s}{D_{max}} = \frac{1}{2}\left[\sqrt{1-\left(\frac{t}{D_{max}}\right)^2} + \sqrt{\left(\frac{D_{min}}{D_{max}}\right)^2 - \left(\frac{t}{D_{max}}\right)^2}\right] \tag{6.2}$$

Figure 6.3 shows a plot of the normalised spacing s/D_{max} versus the ratio D_{min}/D_{max} necessary to obtain the normalised thickness t/D_{max}.

For massive structures formed by layouts of more than two overlapped columns, it is important to know the amount of treated and untreated soil.

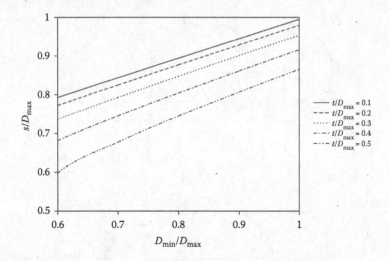

Figure 6.3 Thickness of an element made by two parallel overlapped columns, having different diameters, as a function of diameters and spacing.

A filling ratio may be thus defined as the fraction of the cross-sections effectively cemented by jet grouting

$$F = \frac{A - A_{un}}{A} \tag{6.3}$$

in which A and A_{un} are, respectively, the entire grid area and the untreated area (Figure 6.4). Given a diameter D and a spacing s, this fraction can be easily calculated for the two most typical grid arrays sketched in Figure 6.4. The plot reported in the same figure shows that the equilateral triangular array is more convenient than the square one because, for a given relative spacing S/D, it corresponds to a larger filling ratio F and a smaller overlapping of the columns.

In the previous chapters, it has been repeatedly recalled that columns are far from being perfect cylindrical bodies and that their axes may also diverge from their prescribed alignment. It follows that their true spacing and filling ratios may differ from the ideal values, likely changing along the column axis. Since such differences may be so relevant to affect the performance of the whole jet-grouted structure, the effect of geometrical variability should be carefully considered.

For some of the geometrical factors, like the length of columns and the initial position of perforation, the margins of uncertainty are usually acceptable because it is possible to control them with sufficient accuracy

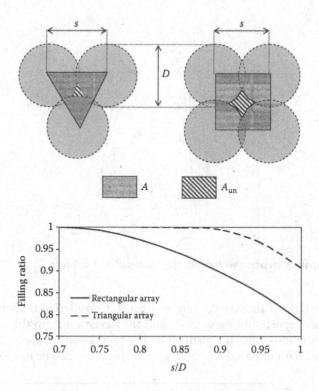

Figure 6.4 Filling ratio of jet-grouted elements arranged with equilateral triangular and square arrays.

(see Chapter 8). As a consequence, unless systematic errors in column production are introduced, it is expected that the prescribed position of the axis head and the design values of the column length will always be respected. For the other relevant geometrical variables, that is, the diameter and the inclination of columns, as well as for the mechanical properties of the jet-grouted material, the uncertainties are usually higher and, as already mentioned, random variability is expected.

The effect of the geometrical variability on the overlapping of adjacent columns can be shown with reference to the critical role played by column inclination. Clearly, for two parallel columns, the distance between their axes is constant, being equal to that imposed at ground level s_o (Figure 6.5). As a consequence, whatever the thickness t of the jet-grouted element (Equations 6.1 and 6.2), its value would also be constant. However, because a deviation from the prescribed position of the axis is possible, the spacing s and, therefore, the thickness t may actually vary along the columns. For vertical columns, being z the distance from the borehole's head,

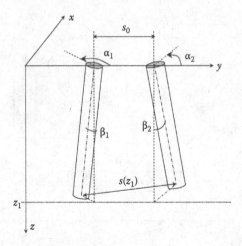

Figure 6.5 Spacing variation with depth s(z) for misaligned columns.

this means that a function $s(z)$ can be defined. Considering the scheme and the symbols reported in Figure 6.5, and the deviation from the verticality of two contiguous columns (defined by the two deviation angles β_1 and β_2 and the two azimuth angles α_1 and α_2), this function can be calculated as

$$s(z) = s_o + z \cdot \sqrt{\tan^2 \beta_1 + \tan^2 \beta_2 - 2 \tan\beta_1 \cdot \tan\beta_2 \cdot \cos(\alpha_1 - \alpha_2)} \qquad (6.4)$$

Figure 6.5 and Equation 6.4 show that $s(z)$ increases linearly with z. The worst condition for overlapping occurs when the two columns are diverging, that is, for $(\alpha_1 - \alpha_2) = 180°$. In this most critical case, the square root of Equation 6.4 becomes equal to $(\tan \beta_1 + \tan \beta_2)$.

By combining Equation 6.1 or 6.2 with Equation 6.4, the thickness $t(z)$ of two overlapped jet-grouted columns can be calculated, provided that the diameters $D_1(z)$ and $D_2(z)$, the azimuths α_1 and α_2 and the inclinations β_1 and β_2 of the two columns are given.

Similar considerations can be made for assemblies of overlapped columns with different grid geometries, although the relevant expressions become formally more complex.

The previous analysis shows that the actual thickness of a jet-grouted element made of two adjacent columns can be defined at the design stage, as a function of depth, if the values of the angles α_1, α_2, β_1 and β_2 are estimated.

In Chapter 4, it has been shown that the variability of the geometrical features (diameter and direction) of a single column can be reasonably estimated by performing statistical analyses of the experimental results and by deriving simple probabilistic models able to simulate variability. The most suited probabilistic distributions are reported in Table 6.4. In Chapter 4, typical values of the scatter parameters have also been given for different soils, which can be adopted if no experimental data are available.

The variability of geometrical factors produces arrays with shapes different from the ideal ones. In the most critical cases, weaker sections or discontinuities capable of affecting the functionality of jet-grouted elements may thus be produced.

In Figure 6.6, some of the most commonly adopted arrays of jet-grouted columns are reported together with an example of the possible configurations

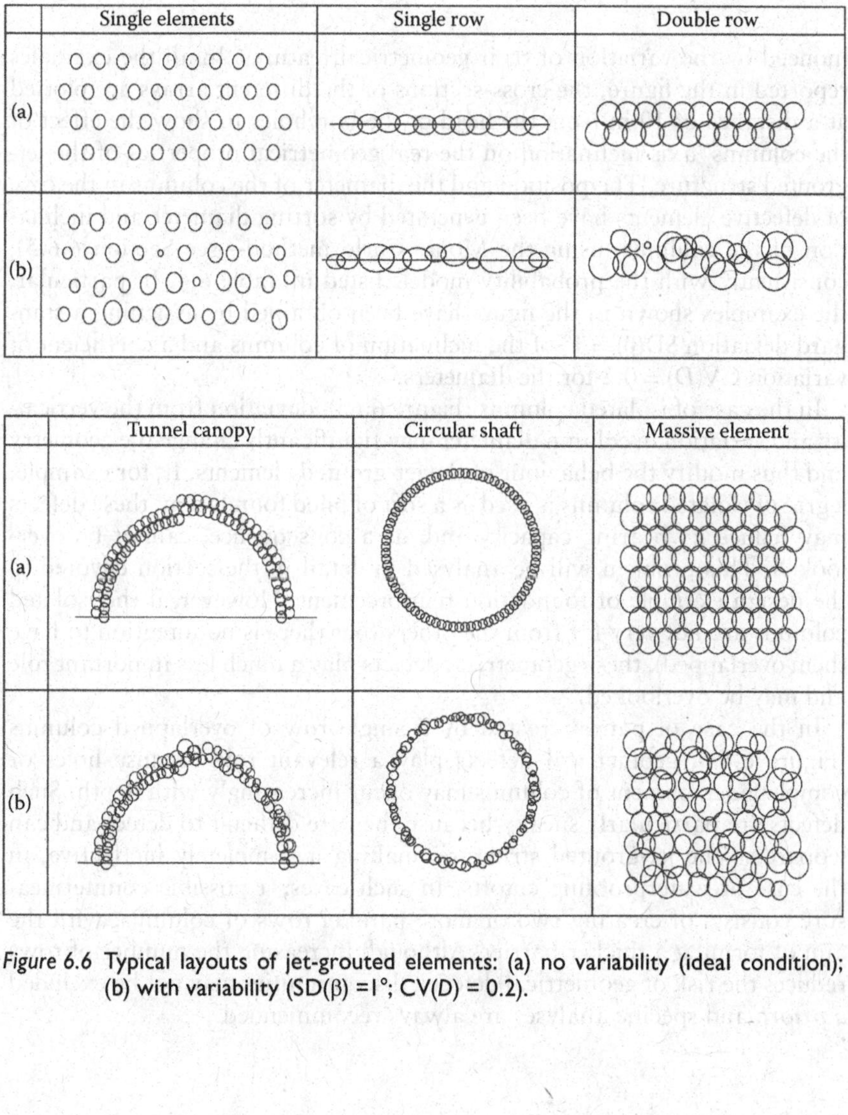

	Single elements	Single row	Double row
(a)			
(b)			

	Tunnel canopy	Circular shaft	Massive element
(a)			
(b)			

Figure 6.6 Typical layouts of jet-grouted elements: (a) no variability (ideal condition); (b) with variability (SD(β) =1°; CV(D) = 0.2).

Table 6.4 Probabilistic distributions suggested for jet-grouted columns

Column characteristics		Probabilistic model	Statistical parameters
Diameter of column		Normal	Mean value Standard deviation
Orientation of column	Azimuth Inclination	Uniform Normal	– Mean value Standard deviation
Strength/stiffness of jet-grouted material		Log-normal	Mean value Standard deviation

induced by the variation of their geometrical features. In all the examples reported in the figure, the cross-sections of the different arrays are plotted at a distance of 10 m from the head of the boreholes to show the effect of the columns' axis inclination on the real geometrical properties of the jet-grouted structure. The position and the diameter of the columns in the case of defective elements have been generated by sorting diameter and inclination of the columns (using the Monte Carlo method—see Section 6.6.5), consistently with the probability models listed in Table 6.4. In particular, the examples shown in the figure have been obtained by assigning a standard deviation $SD(\beta) = 1°$ of the inclination of columns and a coefficient of variation $CV(D) = 0.2$ for the diameters.

In the case of isolated columns (Figure 6.6a), deviation from the verticality and variation in column diameter may significantly change the geometry and thus modify the behaviour of the jet-grouted elements. If, for example, a grid of isolated columns is used as a sort of piled foundation, these defects may reduce its bearing capacity and, as a consequence, cannot be overlooked. This problem will be analysed in detail in the section devoted to the design example of foundation reinforcement. However, if the isolated columns are not very far from the others (but there is no intention to have them overlapped), these geometrical defects play a much less important role and may be overlooked.

In the case of panels created by a single row of overlapped columns (Figure 6.6a), geometrical defects play a relevant role because holes or complete detachment of columns may occur increasingly with depth. Such defects are particularly sneaky because they are difficult to detect and can jeopardise the jet-grouted structure, making it completely ineffective, in the case of waterproofing cutoffs. In such cases, a possible countermeasure consists of creating two or more parallel rows of columns, with the aim of forming a thicker barrier. Although increasing the number of rows reduces the risk of geometrical defects, discontinuities may not be excluded *a priori*, and specific analyses are always recommended.

Variability may become even more critical when the continuity of the jet-grouted structure is fundamental to ensure its static performance. The cases of tunnel canopies or shafts reported in Figure 6.6b clearly represent this situation. In the former example, the canopy is represented by splitting each section in two different halves; the section on the left represents the initial section of the canopy, whereas that on the right represents the geometry at a distance of 10 m from the beginning of perforation. The different dimensions of the two sections depend on the conical shape typically given to the tunnel canopies. In particular, the initial canopy section (left part of the figure) is influenced only by the variation of diameter as there is no influence of axis deviation. The situation becomes more critical at some distance from the initial section (right part of the figure) because of the influence of the canopy opening angle, which is needed to achieve a provisional telescopic lining made by a sequence of partially overlapped canopies. The section dedicated to the design of tunnel canopies (Section 7.3.1) will provide more details on the effect of geometrical defects.

A similar problem may arise during the excavation of a vertical shaft supported by jet-grouted columns, reported in Figure 6.6b and analysed in Section 7.3.2.

Finally, the sketch performed for the massive element reported in Figure 6.6 shows that column axes deviation and random diameter reductions could originate holes which may compromise their waterproofing and/or static function (see Section 7.5).

6.3 DESIGN APPROACHES

With respect to other conventional geotechnical technologies, the design of a jet grouting application is characterised by two specific features. The first one derives from the nature of the jet-grouted material, which is rather different from both natural soils and concrete. It is thus necessary to choose the most convenient constitutive model, with particular regard to the failure criterion. The second important peculiarity consists of the large variability of the geometrical features and mechanical properties of the jet-grouted elements. Thus, a second and fundamental choice is required to account for such inherent variability.

Two alternative failure criteria may be used for the jet-grouted material, based on two different assumptions, as already reported in Section 4.5.3.1.

With the first assumption, the behaviour of the jet-grouted material is still described by the Mohr–Coulomb failure criterion as usually adopted for the untreated soil. Therefore, with this approach, calculations are performed by considering the jet-grouted material as a frictional and cohesive soil of improved quality. This choice is best suited when the percentage of grouted material is relevant (massive treatments). In case not all the volume of soil is grouted, homogenisation techniques may be used to derive the parameters of the overall material (natural soil + jet-grouted material).

With the second assumption, the jet-grouted elements are considered as individual bodies of a completely different material. Therefore, in the calculations, the jet-grouted elements can be treated similarly with concrete structures, considering a Tresca failure criterion.

Whatever the adopted failure criterion, it is necessary to define the geometrical and mechanical properties of the jet-grouted elements. To this aim, as a first step, it is necessary to estimate the characteristic values of the relevant variables (diameter of columns; axes inclination and azimuth; physical, mechanical and hydraulic properties of the jet-grouted material). These characteristic values are intended as cautious estimates of the expected values of the variables. How cautious such estimate should be depends on the quality of the available information (previous experience, field and laboratory data), and this certainly leaves a large degree of subjectivity in the quantification. If a parameter is subjected to random variability and a sufficient number of data is available, its characteristic value could be considered as a (cautious) estimate of the mean value of the parameter.

The second step is the derivation of design values from the characteristic ones. This topic has been extensively treated in the literature (e.g., Fenton and Griffiths 2008) and in the Codes of Practice, but not with specific reference to jet grouting. The following three alternatives may be used with reference to jet-grouted elements to define the design values of the geometrical and mechanical parameters:

- *Deterministic approach*: With this approach, typical in geotechnical engineering and proposed, for example, by Eurocode 7, the uncertainty on the properties of jet grouting is considered by modifying the actions on the structures and the characteristic values of the material properties with partial factors to derive design values. For the material properties of jet-grouted elements, the characteristic values can be derived from the literature or, preferably, be taken from *in situ* measurements (as the mean values), and possible values of the partial factors are suggested in the following paragraphs of this chapter.
- *Semiprobabilistic approach*: If the distributions of the variables are known, the design values can be associated with an acceptable probability of failure and can be computed from probabilistic models. Even in this case, the distributions can be obtained by direct experimental investigations (field trials or laboratory tests) or by using the probabilistic models suggested for jet grouting (see Chapter 4), as

well as typical values of the parameters (e.g., the mean value and the standard deviation of the distributions). Depending on the considered variable, low or high fractiles of the distributions are taken to consider the most risky conditions: in the case of the diameter or the strength of columns the probability of having values lower than the assumed design must be kept as low as possible; in the case of the inclination of the axis of the columns, on the contrary, high fractiles must be considered to reduce the probability of having values higher than the assumed design ones.

- *Probabilistic approach*: If a statistically meaningful sample of the geometrical and mechanical properties of jet-grouted columns is available, a possible alternative is to model the observed distributions with probabilistic functions and to introduce them in a fully probabilistic analysis of the response of the jet-grouted structure. The probabilistic functions of the basic variables can be adopted to compute the distribution of relevant related variables (e.g., the thickness of a jet-grouted element or the axial resistance of a jet-grouted column) to obtain design values of these variables with reference to a prescribed probability of failure. In this case, the solution of the static or hydraulic problem is left to a second calculation stage similar in principle to that described in the semiprobabilistic approach. For more complex problems, where no explicit dependencies can be established between the performance of the structure and the properties of columns, the Monte Carlo method (Haldar and Mahadevan 2000; Fenton and Griffiths 2008) can be used. With this procedure, a large number of structures are generated, and the mechanical or hydraulic performance of each of them is studied. Finally, the probability of failure is quantified by the fraction of unsuccessful cases. Although calculations may be more cumbersome, this fully probabilistic approach has the advantage of considering a more realistic combination of the different variables and of directly relating the performance of the jet-grouted structure to a probability of failure.

Of course, the partial factors, or the probability of failure pertaining to each design approach, should be chosen considering the consequences of an unsuccessful prediction and the role of the considered variable in the overall behaviour of the structure. As an example, an overestimate of the diameter assumes a largely different meaning if the static performance of a structure is ruled by each single jet-grouted column, whose collapse may induce significant consequences, or if the columns represent elements of a massive structure in which local defects have a minor static relevance.

6.4 DESIGN PROPERTIES OF COLUMNS

The deterministic and semiprobabilistic approaches require the selection of design values for the geometrical and mechanical properties of columns. This issue is considered in Sections 6.4.1 (diameter), 6.6.2 (inclination) and 6.6.3 (strength).

The probabilistic approach is discussed in Section 6.4.5.

6.4.1 Diameter of columns

6.4.1.1 Deterministic approach

If direct experimental data are missing, as is rather frequent at the design stage, the characteristic values of the column's diameter D_k (intended as the best possible estimate of the diameter) may be derived by correlations available in the literature (see Chapter 4). If experimental values are available from past experience or field trials, the definition of the characteristic value D_k can be based on a careful analysis of the data.

In both cases, the design value D_d may be obtained by reducing the characteristic value by a partial factor γ_D:

$$D_d = \frac{D_k}{\gamma_D} \tag{6.5}$$

A list of the possible values of γ_D is suggested in Table 6.5, considering the soil heterogeneity (which influences the expected variability of the diameter) and the available experimental information (which influences the ability to predict the mean diameter). In addition, the suggested values of γ_D depend on the kind of jet-grouted structure: in particular, lower values are suggested for massive structures, less sensitive to a possible error in the estimate of the diameter.

Table 6.5 Suggested partial factors γ_D for the diameter of jet-grouted columns

| Application | Available experimental information | γ_D | | |
		Low soil heterogeneity	Medium soil heterogeneity	High soil heterogeneity
Isolated columns, thin structures	Poor	1.10	1.15	1.25
	Good	1.00	1.05	1.10
Massive treatments	Poor	1.05	1.10	1.20
	Good	1.00	1.00	1.05

In all cases, the table indicates that a larger variability should be expected for more heterogeneous soils (see Table 4.9).

6.4.1.2 Semiprobabilistic approach

If a set of measured diameters is available and the experimental data form a statistically representative sample, a design value could be calculated from the normal probabilistic distribution, assigning a probability of failure (P_f) equal to a prescribed fractile of the distribution. Typically, a P_f equal to 5% can be assumed for reference, but as long as no specific value is assigned by codes, different values may be adopted depending on the relevance of the structure under analysis and on the consequences of failure. The design diameter can be computed as follows:

$$D_d = D_{P_f} = D_{mean} \cdot (1 - g(n) \cdot CV(D)) \qquad (6.6)$$

in which D_{mean} and $CV(D)$ are, respectively, the mean and the coefficient of variation of the diameter, computed on the sample of n data, and $g(n)$ is the distribution coefficient defined in Section 4.3.3. A list of values of $g(n)$ for different probabilities P_f and number of available data is reported in Table 6.6.

To make a comparison between the deterministic and semiprobabilistic approaches, the mean value of the diameter can be taken as the characteristic value $(D_{mean} = D_k)$. Then, combining Equations 6.5 and 6.6, the (deterministic) partial factor γ_D can be expressed as

$$\gamma_D = \frac{1}{1 - g(n) \cdot CV(D)} \qquad (6.7)$$

Table 6.6 Distribution coefficient $g(n)$ (Equation 6.6) for the diameter of columns

P_f		$g(n)$		
	5	10	15	20
n 10	1.73	1.34	1.09	0.88
20	1.69	1.31	1.06	0.86
30	1.67	1.30	1.05	0.86
50	1.66	1.29	1.05	0.85
100	1.65	1.29	1.04	0.85

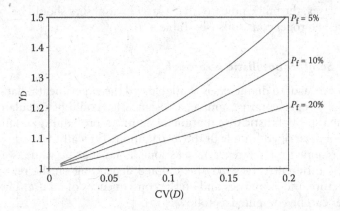

Figure 6.7 Diameter partial factors γ_D computed for different accepted probability of failure P_f and variation coefficients (for a number of available data equal to 20).

Equation 6.7 shows that a rational estimate of γ_D depends on the number of available experimental data n and on the coefficient of variation of the diameter $CV(D)$. Figure 6.7 plots Equation 6.7 for $n = 20$ and three different values of the probability of failure (5%, 10% and 20%). The values of γ_D suggested in Table 6.5 have been calibrated, considering $P_f = 5\%$ and assuming that, in all cases, $CV(D)$, which depends on soil heterogeneity (see Section 4.3.3), is lower than 0.1. Higher values have been measured (see Table 4.8), but are mostly caused by soil stratifications. Obviously, if the number of experimental data n is higher than 20, the deterministic approach would be more conservative, assuming the γ_D values given in Table 6.5, whereas the opposite would happen if $n < 20$.

6.4.2 Inclination of columns

6.4.2.1 Deterministic approach

The second geometrical feature to consider is the inclination (i.e., deviation from the design position) of the column axis, which is fundamental to ensure the effective overlapping of adjacent columns. In verifying the continuity of a jet-grouted structure at the design stage, it is recommended to consider the worst possible condition for the azimuth divergence $\Delta\alpha$ between adjacent columns by assigning the most critical value of α. For two contiguous columns, it should be assumed that $(\alpha_1 - \alpha_2) = 180°$ (see Equation 6.4 and Figure 6.5); for other geometrical arrays, a similarly conservative choice is recommended.

The design value of the deviation β_d should be chosen according to the risk associated with the loss of continuity. In the absence of experimental measurements, it is suggested to adopt a deviation of columns from its theoretical axis $\beta_d = 0.2° - 0.6°$. The largest values of such an interval should be preferred in case of unfavourable conditions, such as low-quality equipment, no measurement of drilling inclination and very heterogeneous and/or coarse-grained soils. Measurement of column inclination performed during field trials may be useful to refine the chosen design value.

6.4.2.2 Semiprobabilistic approach

If a sufficiently large set of measured data, representative of the field conditions, is available, the design value of the deviation angle β_d can be computed by fitting the experimental data with a normal probabilistic distribution (Table 6.4) and by computing the value correspondent to a fractile consistent with the assumed risk. If a 5% probability of being exceeded is assumed, the deviation angle β_d can be computed as

$$\beta_d = g(n) \cdot SD(\beta) \tag{6.8}$$

where $SD(\beta)$ represents the standard deviation of the distribution of the divergences of the column axes from their theoretical position. For the quantification of $g(n)$, the data provided in Table 6.6 can be conveniently adopted, choosing an accepted probability of failure (strictly connected to the assumed risk) and considering the number of experimental observations.

For the azimuth angle α, being its distribution uniform, all values have the same probability of occurring, and as a consequence, the choice of a single design value cannot be but fully deterministic and based on the most unfavourable situation, as previously mentioned and also shown in some of the examples reported in Chapter 7.

6.4.3 Mechanical properties of the jet-grouted material

As discussed in Chapter 4, in most of the usual jet grouting applications, a linear elastic, perfectly plastic model can be adopted to represent its mechanical behaviour, neglecting any tensile strength of the jet-grouted material. The strength of the jet-grouted material is generally expressed by the uniaxial compressive strength (q_u), although, in particular cases, it may be appropriate to adopt the Mohr–Coulomb failure criterion, estimating the shear strength parameters c and φ with laboratory tests.

Assuming a linear behaviour of the material before failure, the value of the Young modulus E can be taken as follows (see Figure 4.25):

$$E = \beta_E \cdot q_u \tag{6.9}$$

The available data suggest that $200 < \beta_E < 700$ (see Table 4.12).

6.4.3.1 Deterministic approach

If no experimental data are available, the characteristic value of the uniaxial strength $q_{u,k}$ may be chosen, following the suggestions reported in Chapter 4.6. If a sufficient number of experimental data are available, $q_{u,k}$ can be rationally estimated by analysing the experimental results, the design value $q_{u,d}$ can be obtained using a partial factor γ_M as

$$q_{u,d} = \frac{q_{u,k}}{\gamma_M} \tag{6.10}$$

with suggested values of γ_M usually larger than 1.5 for all soils. In the case of temporary structures, the value $\gamma_M = 1.5$ can be assumed, whereas larger values should be taken in case of permanent jet-grouted structures. Values of γ_M lower than 1.5 can be considered only if detailed experimental information is available and a low variability is expected. Indications on the suggested values of this coefficient can be found in DIN4093 (2012) and JJGA (2005).

6.4.3.2 Semiprobabilistic approach

When a statistically meaningful sample of experimental data is available, a log-normal probabilistic distribution may be assumed, and the design value of the uniaxial compression strength can be calculated as a function of the assumed probability of failure, with the following equation:

$$q_{u,d} = \exp\left\{\ln(q_{u,mean}) - 0.5 \cdot \ln\left[1 + \left(CV(q_u)\right)^2\right] - g(n) \cdot \sqrt{\ln\left[1 + \left(CV(q_u)\right)^2\right]}\right\} \tag{6.11}$$

where the values of $g(n)$ can be derived as a function of the accepted probability of failure and the number of available data from Table 6.6.

As extensively discussed in Chapter 4, the mean uniaxial compressive strength of a jet-grouted column can be assumed to be equal to the mean value of the distribution of the experimental data obtained on small

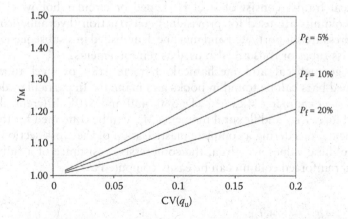

Figure 6.8 Strength partial factors γ_M computed for different accepted probabilities of failure and variation coefficients (the number of available data is taken as equal to 20).

specimens. On the contrary, the coefficient of variation of q_u for the entire column will be significantly lower than that calculated from the laboratory tests on small specimens. In particular, as shown in Chapter 4.5.3.3 (Equation 4.19), the $CV(q_u)$ for the column is approximately one order of magnitude lower than the value computed from the results of laboratory tests. Therefore, considering the experimentally calculated values of $CV(q_u)$ obtained on small-scale specimens (see Chapter 4.5.3.3; Figure 4.27), a maximum value of $CV(q_u) = 0.1$ can be reasonably assumed for the column.

A comparison between the two approaches (deterministic and semi-probabilistic) is reported in Figure 6.8 using Equation 6.7; $q_{u,mean}$ has been considered as the characteristic value, and the partial factor γ_M of Equation 6.10 has been computed for a number of available data $n = 20$ and for different probabilities of failure P_f. The figure shows that, in this case, the previously suggested values of γ_M (≥ 1.5) make the deterministic approach more conservative.

6.4.4 Mechanical behaviour of reinforced columns

The structural performance of jet-grouted columns can be enhanced with the addition of reinforcing elements (generally, steel bars or tubes) pushed into freshly injected columns or placed by perforation (and subsequent grouting) into the hardened jet-grouted material (see Chapter 2.5). Most of the time, these reinforcements are positioned in the centre of the columns, although, in some cases, the reinforcing elements may be located in eccentric positions, aiming to increase the flexural resistance of the column.

Typical frames consist of steel H-shaped or circular hollow elements. If the columns are used for provisional construction (hydraulic isolation, earth pressure support, etc.) and must be demolished in a subsequent phase, fibreglass tubes or rods are also used as reinforcements.

The geometrical and mechanical characteristics of steel tubes and H-shaped bars can be found in books and manuals. In particular, depending on steel characteristics (e.g., Fe 360, 430 and 510), different ultimate normal forces T_{ult} and flexural moments M_{ult} can be computed for the reinforcement, considering a complete plasticisation of the cross-section.

Once these values are given, the compressive resistance of a fully compressed reinforced column can be easily computed as

$$N_{ult} = q_u \frac{\pi \cdot D^2}{4} + T_{ult} \tag{6.12}$$

while in traction $N_{ult} = T_{ult}$.

In case the adherence between the steel element and the jet-grouted material has to be verified, a value of the bonding strength f_{bd} between the steel (or fibreglass) and the surrounding jet-grouted material should be assigned. Typically, f_{bd} can be expressed as

$$f_{bd} = \alpha_{bd} \cdot q_u \tag{6.13}$$

where values of $\alpha_{bd} = 0.05 - 0.07$ are suggested.

Different failure mechanisms and stress distributions may occur in bending depending on the cross-section arrangement, that is, the combination between resistance and the dimensions of the jet-grouted column and the reinforcement. In the example shown in Figure 6.9a, the compressed portion of the jet-grouted column, identified by the shaded area and quantified by the height h, can be determined by finding the angle θ from the following equilibrium equation, written under the assumption of nil tensile strength of the jet-grouted material:

$$(2\theta - \sin 2\theta) = \frac{8 \cdot T_{rf}}{q_u \cdot D^2} = \frac{2 \cdot \pi \cdot T_{rf}}{\chi} \tag{6.14}$$

where χ is the force representing the unconfined compressive strength of the cross-section of the column, calculated as

$$\chi = q_u \cdot \frac{\pi \cdot D^2}{4} \tag{6.15}$$

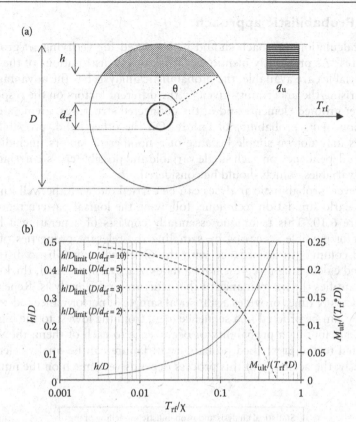

Figure 6.9 Cross-section of a reinforced column (a) and calculation of the compressed section of the jet-grouted material (b).

It is worth noting that Equation 6.14 holds as long as the compressed part of the column (shaded area in Figure 6.9a) does not intercept the reinforced core of the section. This check can be made in Figure 6.9b by comparing the ratio h/D with its limit values expressed by the dotted curves plotted for the different ratios D/d_{rf}. If h/D is smaller than h/D_{limit}, the ultimate bending moment of the reinforced column can be computed on the right axis of the same Figure 6.9b. In case of an opposite response, which occurs when the tensile strength of the reinforcement T_{rf} becomes predominant in comparison with the total compressive strength of the column section χ, compressive and tensile strengths are simultaneously developed in the reinforcement's section. A more refined calculation would be needed to compute the ultimate flexural moment, considering the shape of the reinforcement. However, in this case the ultimate bending moment of the reinforcement can be indicatively assumed as representative of the whole column's section.

6.4.5 Probabilistic approach

What calculation approach should be preferred for verifying jet-grouted structures? As previously mentioned, if statistical distributions of the random variables are available, the probabilistic analysis has the advantage of summarising the uncertainty given by the different factors on the response of the jet-grouted elements and of the connected structures, giving a clear indication of its probability of failure. On the other hand, probabilistic analyses may not be simple because of a number of factors, including a complex dependency on each single variable and possible cross-correlations among variables, which should be considered.

However, probabilistic analyses can be carried out with the well-known Monte Carlo simulation technique, following the logical pattern reported in Figure 6.10. This technique essentially consists of generating a large number of possible scenarios by sampling the relevant properties of jet-grouted columns in accordance with the observed probabilistic distributions and calculating a representative index (e.g., ultimate load, thickness) that quantifies the performance of the structure under analysis. Repeating this process N times provides a statistical sample of performance index values; they can be ordered in a sequence (e.g., from the lowest to the highest values), attributing a probability of occurrence to each of them; the value associated to the prescribed probability of failure can be finally selected. Obviously, the accuracy of this process depends very much on the number

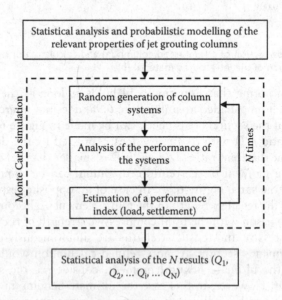

Figure 6.10 Flowchart of probabilistic analysis.

of performed trials because a limited number of simulations could produce a sample not particularly meaningful from a statistical viewpoint. On the other hand, increasing the number of simulations would add robustness to the analysis, producing convergence to a more accurate result, but at the expense of more cumbersome and time-consuming calculations.

A minimum number of trials have to be performed to obtain a prescribed level of accuracy in the estimate. Generally, this number depends on different factors, including the scattering of performance indexes, the accepted probability of failure, the error tolerated in the estimate and the confidence required from the prediction. The literature on this topic (see Fenton and Griffith 2008 or Baecher and Christian 2003 for a more thorough coverage) typically indicates that the number of trials should be of at least hundreds, or possibly thousands, and that variance reduction methods (e.g., importance, antithetic or stratified sampling) can be appropriately adopted to produce a more rapid convergence and to reduce the number of simulations.

With particular reference to jet grouting applications, recent studies (e.g., Modoni and Bzówka 2012) have revealed that estimates with ±5% error and a 90% level of confidence can be obtained by performing 1000 trials, but this number significantly increases if lower errors or higher confidences are sought. In any case, the best way to decide which is the minimum number of simulations needed to have a reliable result is to carry out several Monte Carlo analyses, with increasing the number of trials. When the result becomes insensitive to further increases of the number of trials, the optimum number of trials has been reached. It is worth considering that a large number of simulations can be performed only by implementing automated calculation algorithms.

The previously described probabilistic analysis should be carried out on the whole jet-grouted structure to get the true probability of failure. In the case of large or geometrically complex structures, where calculations could become very cumbersome, it is usually convenient to analyse a representative unit (a mono-, bi- or three-dimensional element, depending on the considered structure). However, it is pointed out that the probability of failure of the representative unit is lower than that of the whole structure and that overlooking this difference may introduce a hidden risk at the design stage. On the other hand, it must be considered that the structure may also tolerate defects (i.e., local failures), provided that they do not affect serviceability and safety. Case-by-case considerations have to be made when particularising the analysis to the different jet-grouted structures.

To show the possible convenience of a probabilistic approach, it is interesting to compare it with the deterministic and semiprobabilistic ones by considering the simple exemplary case of the ultimate normal force N_{ult} in a jet-grouted column. In this problem, two random variables (the diameter D

of the column and the uniaxial compressive strength q_u of the jet-grouted material) combine through the following relation:

$$N_{ult} = q_u \frac{\pi \cdot D^2}{4} \tag{6.16}$$

With a probabilistic approach, combinations of typical values of the coefficients of variation $CV(D)$ and $CV(q_u)$ are considered via a Monte Carlo procedure (Figure 6.10). Random D and q_u values, compatible with the prescribed distributions (normal for diameters, log-normal for uniaxial compressive strength), are generated, and the distribution of limit axial forces N_{ult} (Equation 6.16) is obtained. The value corresponding to the 5% fractile of the distribution have herein been considered as probabilistic estimates of the design ultimate compressive force ($N_{ult,d,prob}$).

With the semiprobabilistic approach, the design values D_d and $q_{u,d}$ correspondent to a 5% fractile have also been calculated from their respective probabilistic distributions, and the correspondent value $N_{ult,d,semiprob}$ has been obtained by substituting D_d and $q_{u,d}$ into Equation 6.16.

The comparison between the two approaches is made by, considering the following reduction factor RF:

$$RF = \frac{N_{ult,d}}{q_{u,mean} \dfrac{\pi \cdot D_{mean}^2}{4}} \tag{6.17}$$

that is, by scaling the $N_{ult,d}$ values obtained with the two previously described approaches with that calculated introducing the mean values ($q_{u,mean}$ and D_{mean}) in Equation 6.17.

Figures 6.11a and b show, respectively, the values of RF obtained with the probabilistic and semiprobabilistic approaches for typical values of $CV(D)$ and $CV(q_u)$.

The comparison between the two plots shows the advantage of performing a probabilistic analysis, where the ultimate load is directly related to a probability of failure and where the uncertainties on each single parameter (D and q_u) are simultaneously combined in a unique variable (N_{ult}). On the contrary, managing the variability of each single variable with partial factors (as in the semiprobabilistic approach) may lead to overconservative results (in this case, the simultaneous assignment of 5% probability of failure to D and q_u obviously corresponds to a lower probability of failure on $N_{ult,d}$).

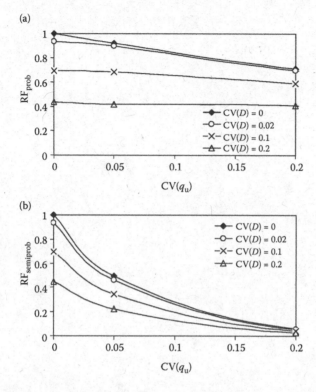

Figure 6.11 Reduction factor RF (Equation 6.17) to calculate the design limit compressive force $N_{ult,d}$ for a probability of failure of 5%: (a) probabilistic approach; (b) semiprobabilistic approach.

It is finally interesting to note that the adoption of the deterministic partial factors given in Table 6.5 ($\gamma_D = 1 - 1.25$) and Section 6.4.3 ($\gamma_M = 1.5$) leads to RF values ranging between 0.53 and 0.67, the minimum value expected for highly heterogeneous soils. The fact that these values fall in the lower range observed in Figure 6.11a confirms that the deterministic approach is rather conservative.

Chapter 7

Design examples

7.1 SELECTED APPLICATIONS

Following the criteria introduced in Chapter 6, the design calculations for some of the most typical jet grouting applications are described in this chapter. The examples herein reported are

- Foundations
- Tunnel canopies and vertical shafts
- Hydraulic cutoffs
- Bottom plugs

The former two applications have a static function, whereas the purpose of the latter two cases is to provide a water barrier to seepage. Jet-grouted foundations are always designed as permanent structures, while jet grouting applications for tunnels and shafts are provisional construction means. Hydraulic cutoffs and bottom plugs may have both provisional and/or permanent functions. Foundations, shafts and canopies are analysed, pointing out the static role of the jet-grouted structures and suggesting practical calculation methods to quantify their resistance against possible collapse mechanisms. The static capacity of these jet-grouted structures is determined, starting from both the geometrical and mechanical characteristics of the jet-grouted elements, and then considering their overall geometrical arrangement and the way multiple elements interact with each other and with the surrounding soil.

In particular, the topic of foundations is dealt with by regarding the two most common geometrical types: jet-grouted blocks and jet-grouted rafts (see Figure 5.2). The latter case is analysed by first examining the behaviour of a single column and then considering the overall performance of the group of columns.

Earth-retaining structures are analysed with reference to curvilinear layouts of jet-grouted columns, which create an arch effect able to turn the soil pressure into compressive stresses acting on the jet-grouted material. This principle is commonly used for the provisional support of tunnel excavations (see Figure 5.18) and of vertical jet-grouted shafts (see Figure 5.8). In these cases, the continuity of the jet-grouted structures provided by the overlapping of adjacent columns is the most important concern, and thus, the analyses mainly focus on this geometrical requirement.

Vertical cutoffs and bottom plugs are the most diffused types of water barriers that may be provided by jet grouting. The continuity of these structures is analysed focusing on their possible geometrical defects.

As already mentioned in the previous chapters, the inherent nonhomogeneity of the geometrical and mechanical properties of the jet-grouted columns plays a key role in the effectiveness of most applications, and therefore, variability has to be considered with particular care. In this chapter, the analysis of the continuity of the jet-grouted structures has been done by means of deterministic, semiprobabilistic and probabilistic methods, considering the experimental evidence introduced in the previous chapters, as well as the probability distributions proposed for the different properties of the columns (diameter, inclination, azimuth, strength). When possible, a comparison of the different approaches has been attempted.

7.2 FOUNDATIONS

The basic principle of foundation reinforcement with jet grouting consists of transferring loads to deeper and more competent subsoil strata, bypassing weaker or more deformable soils. With this goal, two different types of soil reinforcement can be generally obtained: in the first solution, columns are overlapped or very close to each other to form a unique massive body (jet-grouted block) made of jet-grouted material (e.g., Croce et al. 1990); the second solution consists of creating an array of isolated columns forming a system of structural supports similar to a piled foundation (e.g., Modoni and Bzòwka 2012). In both cases, columns may be reinforced with steel frames (bars or pipes) to increase their resistance to axial loads and bending moments.

In the next sections, considerations on the design of these two foundation reinforcement schemes are reported. Ultimate limit state analyses are carried out for both systems subjected to vertical loads, considering collapse mechanisms determined by the failure of the surrounding soil and of the jet-grouted elements.

Particular attention has been paid to the load transfer mechanisms between vertically loaded columns and surrounding soil, which affects the ultimate limit loads and the settlements under working conditions. To this aim, principles currently used for piled foundations have been adapted to jet-grouted columns, starting from the results of available experimental investigations. Extension of these results to arrays of columns has been accomplished in a second step to analyse the overall behaviour of the reinforced foundation.

An analytical method has been finally developed to quantify the response of single reinforced columns subjected to horizontal loads.

7.2.1 Jet-grouted block

The most typical massive reinforcement consists of a block of jet-grouted material with the goal of reducing settlements and providing adequate resistance to vertical and horizontal loads. In the example shown in Figure 7.1, a shallow foundation footing is placed on a block of jet-grouted material. This solution is feasible only for new constructions and would not be reasonable to use as a means of solving foundation problems for existing structures. The working principle of this jet-grouted structure is the transfer of loads to more competent underlying strata or an enlargement of the foundation base to enhance the spreading of loads to the surrounding soil. Such massive reinforcements are sometimes associated with piled foundations to provide additional resistance and enhance the performance of these systems (Miyasaka et al. 1992).

Ultimate limit state analyses must assess the overall stability of the foundation, considering the limit equilibrium of the entire system formed by the jet-grouted block and the surrounding soil, as well as the integrity of the jet-grouted material.

As far as the bearing capacity of the subsoil is regarded, classical solutions may be used. The most critical case pertains to soft, saturated fine-grained soils for which the bearing capacity can be conveniently expressed in terms of total stresses (undrained conditions) using the solution proposed by Terzaghi and Peck (1948) to analyse the block failure mechanism of a group of piles:

$$Q_{\lim} = B_1 \cdot B_2 \cdot (N_c \cdot s_u + \gamma \cdot L) + 2 \cdot L \cdot (B_1 + B_2) \cdot s_u \tag{7.1}$$

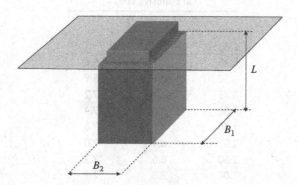

Figure 7.1 Massive foundation reinforcement formed by a jet-grouted block.

in which B_1 is the length, B_2 is the width, L is the depth of the treated soil volume, s_u is the undrained shear strength of the surrounding soil and N_c is the bearing capacity factor reported in Table 7.1.

In granular soils (cohesionless), this general failure mechanism may not be the only one, and local shear and punching collapse mechanisms must be considered as well. A possibility to identify the critical failure mechanism, is given in Figure 7.2 (Vesic 1963), in which the possible failure mechanisms can be derived as a function of the soil relative density D_r and of the ratio between foundation depth (identified as L in Figure 7.2) and width (expressed by the ratio between the area and the perimeter of the footprint).

Alternatively, without making any distinction among the possible different failure mechanisms of the jet-grouted block, a simple formula is provided by Peck et al. (1953) for the case of a strip foundation

$$q_{\lim} = \frac{1}{2} \cdot N_\gamma \cdot \gamma \cdot B_1 + N_q \cdot q \qquad (7.2)$$

where B_1 represents the width of the foundation; γ, the soil unit weight below the foundation plane; q, the overburden stress at the foundation depth; N_γ and N_q, the bearing capacity factors accounting for local shear and punching failure mechanisms (Figure 7.3). Correcting factors (Brinch Hansen 1970; Vesic 1975) can then be introduced to account for different foundation shapes and also to consider other differences from the theoretical scheme (loads inclination and eccentricity, ground inclination), as usual in the calculation of the bearing capacity of shallow foundations.

Table 7.1 Bearing capacity factors for massive reinforcement in cohesive soils

L/B_2	N_c	
	$B_1/B_2 = 1$	$B_1/B_2 > 10$
0.25	6.7	5.6
0.50	7.1	5.9
0.75	7.4	7.2
1.00	7.7	7.4
1.50	8.1	7.8
2.00	8.4	7.0
2.50	8.6	7.2
3.00	8.8	7.4
≥4	9.0	7.5

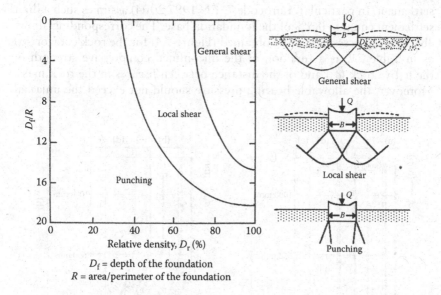

D_f = depth of the foundation
R = area/perimeter of the foundation

Figure 7.2 Failure mechanisms for the massive foundation reinforcement. (Modified from Vesic, A. S., *Bearing Capacity of Deep Foundation in Sand*. Highway Research Board, Record No. 39, 153 p., 1963.)

Sometimes, jet-grouted blocks are created to transfer very high loads to competent rock strata underlying deformable soil layers (Croce et al. 1990). Clearly, in such a case, a global failure mechanism is not possible, and thus, the limit condition is defined with reference to the maximum allowable

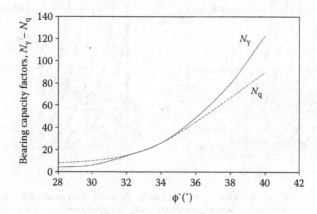

Figure 7.3 Reduced values of the bearing capacity factors accounting for local failure mechanisms. (Modified from Peck, R. B. et al., *Foundation Engineering*: New York: John Wiley & Sons, 401 p., 1953.)

settlement. In particular, Eurocode 7 (EN 1997 2004) assumes such a limit settlement equal to 0.5% of the foundation base. The corresponding allowable bearing pressure can be obtained (Figure 7.4) for the rock categorised as in Table 7.2 as a function of the unconfined compressive strength of the native rock (q_u) and of the distance between fissures in the rock mass. Moreover, the allowable bearing pressure should not exceed the uniaxial

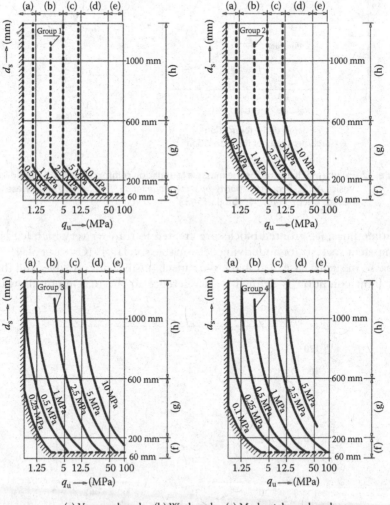

(a) Very weak rock, (b) Weak rock, (c) Moderately weak rock,
(d) Moderately strong rock, (e) Strong rock

Figure 7.4 Bearing capacity of foundations resting on rock. (From EN 1997-1, *Eurocode 7: Geotechnical Design.* European Committee for Standardization: 170 p., 2004.)

Table 7.2 Rock classification for bearing capacity calculation with
Figure 7.4 graphs

Group	Type of rock
1	Pure limestones and dolomites–carbonate sandstones of low porosity
2	Igneous–oolitic and marly limestones–well-cemented sandstones
	Indurated carbonate mudstones–metamorphic rocks, including slates and schists (flat cleavage/foliation)
3	Very marly limestones–poorly cemented sandstones
	Slates and schists (steep cleavage/foliation)
4	Uncemented mudstones and shales

compressive strength of the rock if joints are tight or half of this value if joints are open.

The integrity of the jet-grouted block should be verified by assessing that the applied stresses do not cause failure mechanisms and that plasticisation does not affect significant portions of the jet-grouted block. To this aim, the distribution of compressive stress at the base of the footing of sides B_1 and B_2 can be computed as in Figure 7.5, in the hypothesis of nil tensile

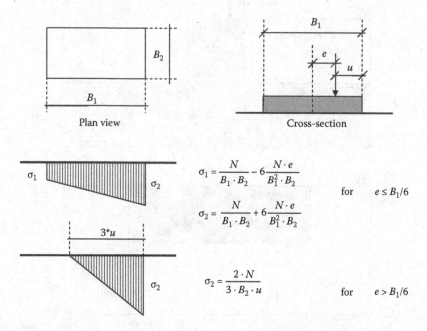

$$\sigma_1 = \frac{N}{B_1 \cdot B_2} - 6 \frac{N \cdot e}{B_1^2 \cdot B_2}$$

$$\sigma_2 = \frac{N}{B_1 \cdot B_2} + 6 \frac{N \cdot e}{B_1^2 \cdot B_2}$$

for $e \le B_1/6$

$$\sigma_2 = \frac{2 \cdot N}{3 \cdot B_2 \cdot u}$$

for $e > B_1/6$

Figure 7.5 Distribution of compressive stresses given on a rectangular footing by eccentric load.

resistance at the contact between the jet-grouted material and the underlaying rock mass.

Then, the assessment of the block integrity can be conveniently performed with numerical analyses to figure out possible failure mechanisms. Two possible failure mechanisms are shown in Figure 7.6. The first one (Figure 7.6a) pertains to the case of a rigid block of jet-grouted material created to bypass a layer of weak soil and to transfer the loads to a deeper bedrock, showing the formation of a sliding wedge in the upper part of the block. In the second example (Figure 7.6b), a jet-grouted block is formed to enlarge the basis of a shallow foundation resting on a deformable soil; for high loads or poor mechanical properties of the jet-grouted material, plasticisation may affect the central portion of the block with the formation of two subvertical cracks near the footing border and a punching failure mechanism.

In these analyses, it is important to cautiously assign the properties of the jet-grouted material accounting for the possible defects occurring at the construction stage or to use homogenisation techniques if jet grouting is not extended to the whole block, which may then include some untreated

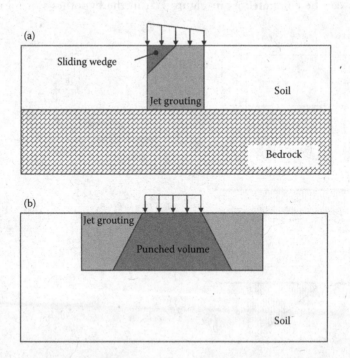

Figure 7.6 Possible failure mechanisms within a massive jet-grouted foundation: (a) sliding wedge; (b) punching.

parts. The methods suggested in the previous chapters may be conveniently adopted to estimate the filling ratio and to assign a design value of the uniaxial compressive strength.

Although continuity is not strictly required for jet-grouted blocks, significant concentration of stresses may occur in the jet-grouted portions where overlapping of columns is incomplete. One possible strategy to analyse this aspect consists in fixing a position of the perforation axes, assuming design values (with the deterministic or semiprobabilistic methods defined in Chapter 6) of the diameter and the inclination of columns, and in calculating the filling ratio with the diagrams of Figure 6.4. A reduced strength can be finally calculated for the equivalent continuous structure by multiplying the design strength assigned to the material for this filling ratio.

Another possibility could be to evaluate the resistance of these jet-grouted structures by means of probabilistic analyses, where the variability of all factors (diameter and inclination of columns and strength of jet-grouted material) is simultaneously considered.

Finally, an analysis of settlements should be performed to check their compatibility with the superposed structures. These calculations can be carried out with the methods currently adopted for concrete foundations, including numerical codes. In this case, the jet-grouted material can be simulated as a linear elastic, perfectly plastic material, being not necessary to introduce more complex constitutive models.

7.2.2 Jet-grouted raft

Foundation reinforcements made of arrays of isolated jet-grouted columns (jet-grouted rafts) may be adopted for new buildings (e.g., Perelli Cippo and Tornaghi 1985; Falcao et al. 2001; Saglamer et al. 2001; Davie et al. 2003) or for the underpinning of pre-existing structures (e.g., Shibazaki and Ohta 1982; Maertens and Maekelberg 2001; Popa 2001). In all cases, the goal of jet-grouted columns is to increase the bearing capacity of the foundation and to reduce the settlements of the superstructure under working loads. The working mechanisms of this kind of jet-grouted structure are similar to those of piled foundations (Viggiani et al. 2011), that is, the load is transferred to the surrounding soil partly from the tip and partly from the lateral surface of the columns.

7.2.2.1 Soil–column interaction

To assess the global performance of the jet-grouted raft, it is important to quantify the interaction between columns and the surrounding soil or, more precisely, to determine how lateral interface stresses and end-bearing load are mobilised with settlements. This aspect, clearly correlated to the geometry of columns, has been investigated with full-scale experiments by different authors (Cicognani and Garassino 1989; Garassino 1997; Maertens and Maekelberg 2001; Bustamante 2002; Bzówka 2009), who generally agree that high loads can be transferred to the surrounding soil from the lateral surface of columns.

In an attempt to compare the performances of different types of columnar reinforcements (nondisplacement piles [NDP], displacement piles [DP], continuous flight auger piles [CFAP] and jet-grouted columns [JGC]), Modoni et al. (2012) have collected the results of different axial loading tests on columns created in sandy soils. To isolate the effect of the installation technique, each measured limit load has been scaled by the product of the mean cone penetration test (CPT) resistance (q_c) measured along the column length with the total surface area (A) of the column (which is the sum of the base and lateral surfaces). In Figure 7.7, the ratio $Q_{lim}/(A \cdot q_c)$ is plotted versus the column slenderness ratio (L/D); the results are grouped considering four different kinds of columnar reinforcements (NDP, DP, CFAP and JGC). According to these results, it seems that jet grouting columns behave similarly to the NDP piles, being less effective in terms of the combination of limit load and slenderness in comparison with more effective installation techniques (displacement and CFA piles). However, the use of jet grouting becomes convenient when columns of limited depth and large diameter are created, that is for the most typical geometrical layouts of jet-grouted rafts. In this case, the bearing capacities are proven to be satisfactory.

The good performance of jet-grouted columns can be explained considering their noncylindrical shape that, as shown in Chapter 4, is due to the random variation of diameters with depth. This irregular shape, artificially

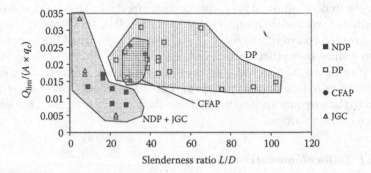

Figure 7.7 Ultimate load of different columnar reinforcements. (Modified from Modoni G. et al., Load-settlement responses of columnar foundation reinforcements. *Proceedings of the ISSMGE–TC211 International Symposium on Ground Improvement 3,* Brussels, Belgium, May 31–June 1, 2012: pp. 491–503, 2012.)

created in underreamed bored piles (Mohan et al. 1967), enables a better transfer of the loads to the surrounding soil. Modoni and Bzówka (2012) examined this effect by back analysing the results of axial loading tests performed on jet-grouted columns in sandy materials and by calibrating a finite element model to derive the load transfer curves along the lateral surface and the lower tip of columns. They found that the ultimate vertical stresses τ_L transferred from the lateral surface of jet columns are comparatively higher than those given for nondisplacement piles by Wright and Reese (1977) (Figure 7.8) and attributed this effect to the funnel shape of columns.

These results are in a good agreement with the values provided by Bustamante (2002), who performed a series of axial loading tests on columns equipped with removable extensometers to separately compute the load fractions transferred to the surrounding soil from the lateral surface and from the tip of columns. Based on these results, Bustamante provides the four charts reported in Figure 7.9, in which the lateral resistance τ_L and the end-bearing capacity p_L are reported versus the penetration resistance, expressed in terms of either the cone tip resistance q_c or the blow count N_{SPT}.

These values can be used to estimate the ultimate limit axial load Q_{ULS} of a single column:

$$Q_{ULS} = P_{ULS} + S_{ULS} \tag{7.3a}$$

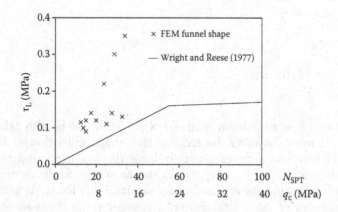

Figure 7.8 Influence of funnel shape on the lateral resistance of jet-grouted columns in cohesionless soils. (From Modoni, G. and J. Bzówka, *ASCE Journal of Geotechnical and Geoenvironmental Engineering* 138(12): pp. 1442–1454, 2012.)

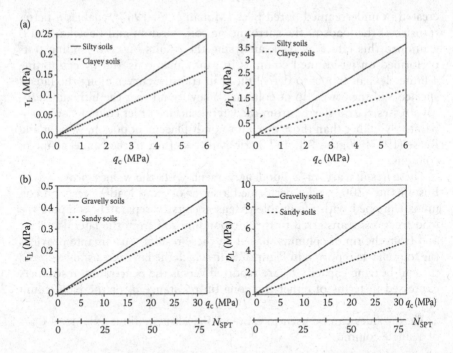

Figure 7.9 Ultimate shear stress at the column–soil lateral interface (τ_L) and end-bearing capacity (p_L) of jet grouting columns in cohesive (a) and cohesionless (b) soils.

$$P_{ULS} = p_L \cdot \pi \frac{D_{base}^2}{4} \tag{7.3b}$$

$$S_{ULS} = \pi \int_L D \cdot \tau_L \cdot dz \tag{7.3c}$$

where a variation of column diameter with depth can be also taken into account. It must, however, be recalled that the mobilisation of loads at the tip of columns requires generally large displacements and thus, for large columns, where P_{ULS} represents the most significant contribution, large settlements may be attained under serviceability loads. As also found on other types of piles, this observation suggests the assessment of the performance of the jet-grouted raft with reference to serviceability limit conditions.

The development of settlements with the applied loads for jet-grouted columns can be seen in Figure 7.10, where 10 different load-settlement

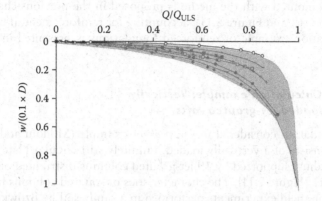

Figure 7.10 Mobilisation of axial loads on jet-grouted columns in cohesionless soils.

curves experimentally obtained on different soils are compared (Modoni et al. 2012). In this figure, a common limit condition has been assumed for all cases as the one giving a settlement equal to 10% of the average column diameter, and thus, the settlements and the loads of each curve have been scaled with reference to their respective values, obtained by extrapolation at this level of settlement (Chin and Vail 1973). If, for instance, a safety factor equal to 2 is taken ($Q/Q_{lim} = 0.5$), settlements ranging between 0.003 and 0.015 of the average column diameter are expected.

7.2.2.2 Column resistance

Another relevant problem to be considered is represented by the possibility of exceeding the strength of the jet-grouted material. This failure mechanism, which is not usually considered for piled foundations, becomes fundamental for jet-grouted columns because of the lower resistance of the jet-grouted material in comparison with concrete, combined with the stress increments caused by the local reductions of diameters typically affecting jet columns. Its relevance on the behaviour of jet-grouted rafts is clearly demonstrated by a series of field and laboratory full-scale tests performed by Maertens and Maekelberg (2001), who observed the sudden structural collapse in the upper part of axially loaded columns, well before the failure of the surrounding soil.

Safety calculations with regard to this issue can be performed by computing the distribution of loads on the top of each column of the jet-grouted structure and by comparing these values with the axial resistance of the single columns. The former calculation can be accomplished by adapting to jet grouting columns the analytical methods defined for piled foundations (e.g., Viggiani et al. 2011). The ultimate axial forces of jet grouting columns can be computed with the methods proposed in the previous chapter (see Equation 6.12 and Figure 6.11), accounting for reinforcement, if necessary. A calculation example of reinforced foundations is presented in the next section.

7.2.2.3 Calculation example: Vertically loaded jet-grouted rafts

The foundation considered in the present example (Modoni and Bzówka 2012) consists of a vertically loaded, infinitely stiff circular plate having a 6.3-m radius, supported by 19 jet-grouted columns distributed on concentric circles (Figure 7.11). The characteristics of soil and columns have been taken from field experiments performed in a sandy soil by Bzówka (2009). In this example, a constant unit cone tip resistance $q_c = 10$ MPa has been

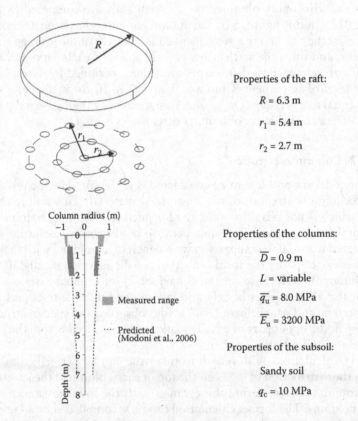

Properties of the raft:

$R = 6.3$ m

$r_1 = 5.4$ m

$r_2 = 2.7$ m

Properties of the columns:

$\overline{D} = 0.9$ m

$L = $ variable

$\overline{q_u} = 8.0$ MPa

$\overline{E_u} = 3200$ MPa

Properties of the subsoil:

Sandy soil

$q_c = 10$ MPa

Figure 7.11 Vertically loaded raft supported by jet-grouted columns. Geometric scheme and material properties.

assigned to the soil, and the columns have been modelled with the shape reported in Figure 7.11, with a mean value of the diameter of 0.9 m, a reduction of diameter with depth and different possible lengths. The mean values of the uniaxial compressive strength and of the Young modulus of the jet-grouted columns are, respectively, 8 and 3200 MPa.

In the numerical model (see Modoni and Bzówka 2012 for details), the load settlement response of single columns has been calculated with the load transfer curve method, calibrated on the results of available loading tests. The distribution of loads on each column has been computed considering the interaction between contiguous columns and assuming a linear soil reaction beneath the plate, with a subgrade reaction factor $k = 3968$ kN/m^3 computed with the method proposed by Schmertmann (1970). Assessment of the integrity of columns is made at all steps of calculation by comparing the axial load acting at any depth of each column to the resistance calculated using Equation 6.12.

The positive effect of jet-grouted columns, in terms of settlement reduction and increase of maximum allowable loads, can be seen by comparing the load-settlement response of the foundation without columns (Schmertmann 1970) to that with columns of different lengths (Figure 7.12a).

In this example, a limit load Q_{lim} has been defined as the one producing a settlement w_{lim} equal to 5% of the mean diameter of columns ($w_{lim} = 46$ mm; see Figure 7.12a). With reference to such a limit load, Figure 7.12b shows the positive effect of increasing column length. However, the figure shows that, for unreinforced columns, it is useless to have columns longer than 22 m. This effect occurs because, in this example, the load $Q = 120$ MN (see Figure 7.12b) produces a compressive stress in the uppermost part of the columns equal to the compressive strength of the jet-grouted material, and thus, structural failure occurs within the columns. The bearing capacity can be increased by adding a reinforcement to the columns (in this example, HEB 240 steel).

The optimum column length can be found as the one giving the simultaneous attainment of the two limit conditions ($w = w_{lim}$ and structural failure of columns), that is, the first point of the final horizontal part of the curves reported in Figure 7.12b. A further increase of column length beyond this value would be useless. In the reported example, this condition occurs for unreinforced columns when the length is 22 m ($Q_{lim} = 118$ MN) and for reinforced columns when the length is 40 m ($Q_{lim} = 166$ MN).

7.2.2.3.1 Probabilistic approach

In the calculations discussed above, the random variability of column properties has not been considered, and the design values have been deterministically assigned. The same example has then been analysed with the probabilistic approach, using the Monte Carlo simulation technique (see

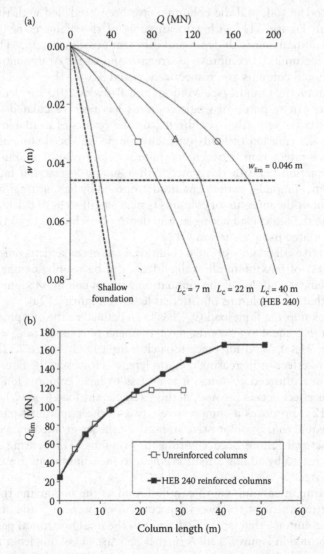

Figure 7.12 Load settlement curves of the foundation with and without columns (a) and influence of the length of columns on the limit load (b). The horizontal dotted line in (a) defines the reference settlement (5% of mean diameter) used to calculate Q_{lim}.

Chapter 6.4 and Figure 6.10), to investigate the influence of properties variability. The diameter D of each column has been assumed as normally distributed around the values adopted in the previous deterministic analysis (i.e., assuming the profile of Figure 7.11); a log-normal distribution has been assigned to the strength q_u of the jet-grouted material, considering

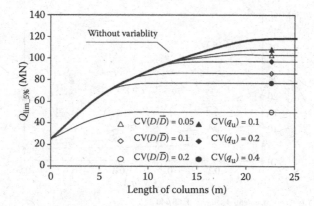

Figure 7.13 $Q_{lim\_5\%}$ for unreinforced columns of different lengths.

the mean value $\bar{q}_u$ equal to 8 MPa, whereas the coefficients of variation $CV(D/\bar{D})$ and $CV(q_u)$ have been varied as reported in Figure 7.13. In the figure, the role of each coefficient of variation has been separately investigated (i.e., q_u has been kept equal to its mean value, whereas varying $CV(D/\bar{D})$ and D has been kept equal to $\bar{D}$ while varying $CV(q_u)$).

The procedure has been applied by running 1000 simulations for each combination of factors $[CV(D/\bar{D}), CV(q_u)]$, by calculating for each of them the limit load Q_{lim} corresponding to an allowable settlement of the foundation of 46 mm (consistent with the assumption made in the deterministic calculation) and by selecting the Q_{lim} value associated to a probability of failure equal to 5%. Figure 7.13 reports the $Q_{lim\_5\%}$ for unreinforced columns of different lengths, clearly showing that the variability of diameter and strength reduces the bearing capacity. Lower values of the optimum column length are thus obtained with the probabilistic calculation. The most important practical consideration on these results is that the improvement of foundations with jet-grouted columns can be effective but is bounded by the limited strength of the jet-grouted material (Figure 7.14a, b) as well as by the variability of strength and diameter. However, a significant improvement in jet-grouted column performance can be obtained by adding reinforcing frames, which increase the structural strength.

7.2.2.4 Laterally loaded single column

7.2.2.4.1 Probabilistic approach

The ultimate lateral resistance (horizontal bearing capacity) of jet-grouted columns (Modoni et al. 2008b) can be evaluated by adopting the theory originally formulated for single laterally loaded piles by Broms (1964a,b).

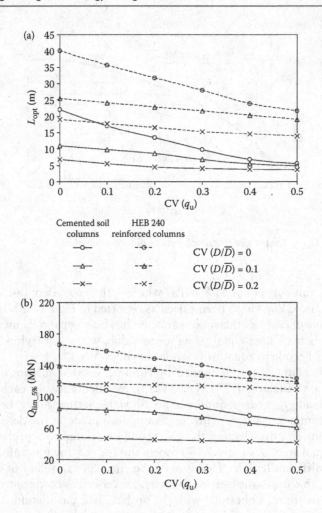

Figure 7.14 Effect of jet grouting variability on the optimum length of columns (a) and on the limit load of the foundation (b).

According to this theory, which considers a rigid–perfectly plastic behaviour of both the soil and the pile (i.e., the column), the horizontal pile–soil interaction pressure $p(z)$ depends on soil strength and column diameter via the equations:

$$p(z) = 9 \cdot s_u(z) \cdot D \quad \text{cohesive soils} \tag{7.4a}$$

$$p(z) = 3 \cdot k_p \cdot \gamma \cdot z \cdot D \quad \text{frictional soils} \tag{7.4b}$$

where D is the diameter of the column; s_u, the soil undrained shear strength; k_p, the passive earth pressure coefficient; γ, the soil unit weight (which becomes γ' if the soil is saturated); and z, the depth from ground level. Although it is, nowadays, well known that the Broms solution for frictional soils is very conservative (e.g., Viggiani et al. 2011), it is still largely adopted in practice and will, therefore, be used in this section. The structural strength of the jet-grouted column can be considered via its ultimate bending moment M_{ult}, calculated with the procedure defined in Chapter 6.4.4.

The horizontal bearing capacity H_{lim} of the single column is then calculated by imposing limit equilibrium conditions for all possible failure mechanisms, assuming the appropriate distribution given by Equation 7.4 for the pile–soil limit interaction pressure.

The example here reported (Figure 7.15) describes the case of a head-fixed, irregularly shaped and nonhomogeneous jet-grouted column. In the calculations, the variability has been considered by dividing the column into n parts, each having length $l_i = L/n$ (L being the column length), a diameter D_i and an ultimate bending moment $M_{p,i}$. Three mechanisms are considered (Figure 7.15): the horizontal translation of the foundation system without structural failure (usually defined as 'short' column mechanism), the creation of a single plastic hinge in the column ('intermediate' column) and the creation of two plastic hinges ('long' column).

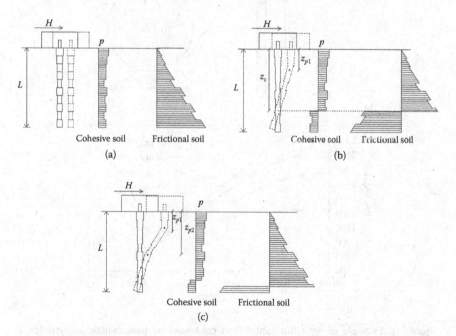

Figure 7.15 Possible failure mechanisms of a head-fixed horizontally loaded irregular column: (a) 'short', (b) 'intermediate' and (c) 'long' columns.

For a short column, the solution is found by imposing the equilibrium to translation in the horizontal direction. For intermediate and long columns, rotational equilibrium relations must be added to calculate H_{lim}, considering that, in the plastic hinges, the ultimate bending moment M_{ult} is applied. This latter solution is not as straightforward as for homogenous and cylindrical piles because the ultimate bending moment varies with depth z, being, therefore, $M_{ult}(z)$. For each of these two mechanisms, the horizontal limit load has been calculated as the minimum among those corresponding to all possible positions of the plastic hinges. Finally, the minimum value of H_{lim} among the three ones calculated considering the mechanisms of Figure 7.15 is the horizontal bearing capacity of the column.

Obviously, the value of H_{lim} largely depends on the assumed distributions $D(z)$ and $M_{ult}(z)$ again. A large number of simulations have been carried out with the Monte Carlo procedure for the case of a fixed-head column having

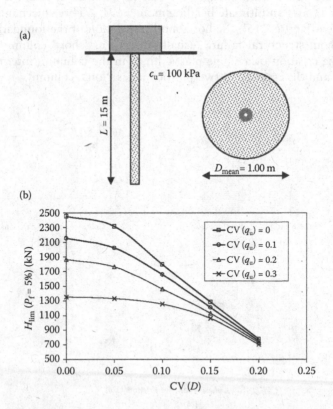

Figure 7.16 Example of limit horizontal load calculated by probabilistic analysis for a fixed-head column in cohesive soil: (a) set-up of the column; (b) results of calculation.

$L = 15$ m, immersed in a cohesive soil having $s_u = 100$ kPa (Figure 7.16a). As usual, the column diameters have a normal probabilistic distribution with a mean value $D_{mean} = 1.0$ m and different coefficients of variation CV (0–0.20). The mean uniaxial compressive strength q_u of the jet-grouted material is distributed with a log-normal probabilistic function, with a mean value $q_{u\_mean} = 5.0$ MPa, and coefficients of variation in the range 0–0.3. A reinforcing steel pipe having an external diameter of 100 mm, a thickness of 12 mm and a tensile strength of 360 N/mm² has been inserted into the column. The results reported in Figure 7.16b refer to a tolerated probability of failure equal to 5%.

The case of an ideal cylindrical and homogeneous column ($CV(D) = 0$, $CV(q_u) = 0$) obviously results in the largest possible value of H_{lim} (2452 kN). Variability in the diameter and mechanical properties of the column leads to a significant reduction in the horizontal bearing capacity, which, in the reported example, can be approximately 30% of the maximum value pertaining to the ideal column. For horizontally loaded columns, the figure also shows the predominant role played by the variation of diameter due to its large influence on the ultimate bending moment $M_{ult}(z)$.

7.3 TUNNEL CANOPIES AND VERTICAL SHAFTS

The use of jet grouting to create earth-supporting structures has greatly increased in recent years. At the design stage, however, there is still a relevant degree of uncertainty, mainly because of the intrinsic defects of jet-grouted columns and the nonhomogeneities of the treated soil. In this section, specific reference will be done to the design of two kinds of peculiar structures for which jet grouting has proven very effective: provisional tunnel support (Figure 7.17a, b) and provisional support of large shafts (Figure 7.17c). These jet-grouted structures have the common feature that the soil

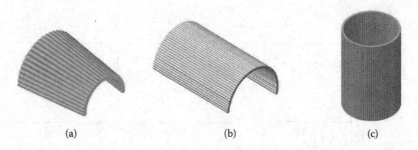

(a)   (b)   (c)

Figure 7.17 Typical curved jet-grouted supporting structures: tunnel canopies with a frustum of cone shape (a) and with a cylindrical shape (b) and vertical shafts (c).

supporting action is mostly committed to the arching effect provided by the overlapping of adjacent columns. In fact, thanks to their curved arrangement, these structures are able to take up the earth pressure by developing a distribution of circumferential compressive stresses, compatible with the characteristics of the jet-grouted material, which significantly helps the static response reducing the longitudinal bending moments. Although the flexural resistance of columns may be improved by inserting reinforcing longitudinal elements (generally steel bars or tubes) (see Figure 6.9), discontinuities of these structures must be carefully avoided, and design analyses must necessarily be carried on with this purpose.

7.3.1 Jet-grouted canopies

In this section, a possible design approach for jet-grouted structures used as temporary supports in tunnelling is considered. Such structures, usually called 'canopies', have been described in some detail in Chapter 5.5 and are an excellent means in soils with low or no cohesion to supply temporary support ahead the front to allow excavation prior to the installation of the final lining. The typical iterative construction sequence, described in Chapter 5.5, is recalled in Figure 7.18.

To allow the construction of subsequent canopies, keeping a constant cross-section area for the tunnel to be realised, these curved structures are made of slightly diverging columns. Therefore, the canopies have the shape of a three-dimensional frustum of cone, whose cross-section area gets larger along the tunnel axis for each span. The resulting jet-grouted structure is a sequence of partially overlapped vaults (Figure 7.19), each having a length and an opening angle assigned at the design stage. Typically, a minimum overlapping of 2–3 m between adjacent vaults is assigned, whereas the inclination angle β obviously depends on the column diameter D and has to be assigned to allow the construction of the subsequent canopy, once the canopy overlapping distance a and the span y_{max} are given (Figure 7.20).

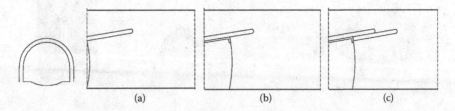

(a) (b) (c)

Figure 7.18 Construction sequence of a jet-grouted canopy: (a) first span; (b) excavation, shotcreting and positioning of steel ribs; (c) iteration of the process.

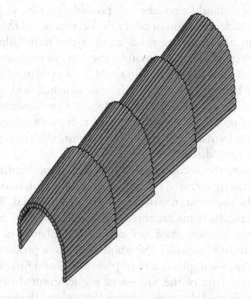

Figure 7.19 Example of a sequence of partially overlapped jet-grouted canopies.

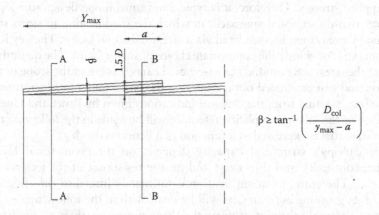

Figure 7.20 Longitudinal section of the canopy sequence, with an indication of the minimum opening angle β necessary to allow a mutual overlapping length a of the canopies.

In Chapter 5, the following three different failure mechanisms of jet-grouted canopies were identified (see Figure 5.20):

- Face failure
- Tip failure
- Lining failure

Design analyses should consider all possible factors of uncertainty to exclude with a sufficient margin of safety the occurrence of these mechanisms.

The most typical countermeasure against the first failure mechanism consists of limiting, as much as possible, the inward movement of the excavation front. This goal is usually obtained by a system of anchoring bars that tie the soil behind the face to the undisturbed soil portion located at some distance from the excavation front. Dimensioning the anchors basically consists of determining the number, type and length of the bars. In this failure mechanism and in its assessment, the resistance of the jet-grouted canopy is not directly involved.

On the contrary, the second and third failure mechanisms sketched in Figure 5.20 are strictly dictated by the distribution of internal stresses within the canopy and by the resistance of the jet-grouted material. The phenomena involved in these mechanisms are three-dimensional and far from having any kind of symmetry (neither axial nor plane). Calculation of stresses in the canopy should carefully consider the whole tunnelling process (i.e., jet grouting execution, excavation, placement of provisional and final linings).

However, a great part of the success of the jet-grouted canopy technique relies on the continuity of the cross sections throughout the longitudinal development of the canopy and on the ability of the jet-grouted material to support the applied stresses. Therefore, it is typical in tunnel lining design to adopt a simpler two-dimensional approach, in which the modification in stress state caused by excavation is considered via a stress reduction factor. The key issue in doing so is obviously the appropriate choice of such a factor. The quantification of the stress state around the jet-grouted canopy is out of the scope of this section and can be carried out considering the large amount of literature on tunnelling, starting from the classical indications given by Panet and Guenot (1982). For the sake of simplicity, reference will be made in the following to a stress state that is expressed as a function of a fictitious depth z.

The canopy's structural capacity depends on the geometrical layout (shape, continuity and thickness) and on the resistance of the jet-grouted material. Therefore, its quantification is intimately linked to the effectiveness of jet grouting activities, as will be dealt with in the following.

From a structural point of view, the three-dimensional jet-grouted canopy can be conveniently modelled as made of longitudinal elements and cross-sectional arches (see Figure 7.21).

Such virtual arches exist as long as a full overlapping of the columns is achieved in the cross-section. The supporting reactions of the arches on the longitudinal elements are possible because of the in-plane stiffness of arches themselves and can be modelled, for instance, as actions given by springs (having the stiffness of the arches). With this scheme, longitudinal elements lay on a continuous field of springs with decreasing stiffness (because the arch diameter increases along each span). If the overlapping of columns is perfect (i.e., the canopy does not have holes), arches are essentially loaded

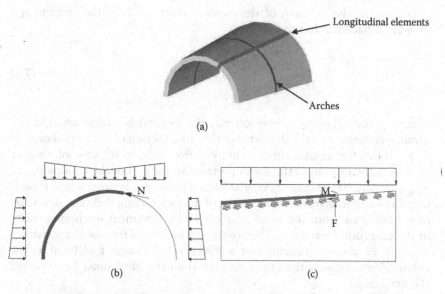

Figure 7.21 Structural idealisation of a jet-grouted canopy (a), modelled with superimposed arches and longitudinal elements. Arches (b) may be considered as
springs supporting the longitudinal elements (c).

in compression, whereas the longitudinal elements are subjected to very
modest bending moments.

Within this modelling scheme, similar to the one adopted in the analyses
of tubes, the bearing capacity of the jet-grouted structure depends on the
load capacity and in-plane stiffness of the arches, as well as on the strength
of the longitudinal beams.

If full overlapping among columns is ensured throughout the whole span,
and the stiffness of the supporting arches is large enough, the beam-like
behaviour can be overlooked, and the simpler bidimensional plane strain
scheme of a series of compressed adjacent arches can be considered in the
analysis. With such a simplification, a properly constructed canopy behaves
as a series of constrained arches loaded from the surrounding soil both
vertically and horizontally.

However, even in the ideal case of a canopy made of perfectly cylindrical
columns exactly positioned along the prescribed design axes, the thickness
of the jet-grouted structure decreases along the tunnel axis because of the
canopy opening angle β (see Figure 7.20). The thickness $t(y)$ of the ideal
arch can be easily found at any abscissa y based on geometrical considerations, once the column diameter (D) and the spacing (s) are known. In
particular, for two adjacent columns (see Equation 6.1):

$$t(y) = \sqrt{D^2 - s(y)^2} \qquad (7.5)$$

Because of the opening of the canopy along its axis, the spacing $s(y)$ between adjacent columns is

$$s(y) = s_0 \cdot \frac{\left(D_{\text{tunn},0} + 2 \cdot y \cdot \tan(\beta)\right)}{D_{\text{tunn},0}} \tag{7.6}$$

where s_0 and $D_{\text{tunn},0}$ are, respectively, the spacing between adjacent jet-grouted columns and the diameter of the cross section of the tunnel canopy at $y = 0$ (beginning of a single canopy). Obviously, in the case of absence of defects the minimum thickness pertains to the end of the canopy ($t_{\min} = t(y_{\max})$, where $s(y)$ has the maximum value). However, because of the typical defects of columns (random variation of diameter along the axis, deviation of column axis from the ideal position and mechanical nonhomogeneity of the jet-grouted material), the continuity and regularity of the structure may not be always granted, and a structural thickness t (defined as the columns' overlapping thickness) smaller than the ideal must be expected (Figure 7.22).

If the condition of overlapping among columns is not met in one of the cross-sections of the canopy its stability should be analysed by also looking to the flexural behaviour of columns in the longitudinal direction. In fact, the bed of springs supporting the columns would be incomplete, with missing springs in the sections where the arch is not complete. This condition would result in the necessity of using longitudinal reinforcement to compensate the non-tensile behaviour of jet-grouted material. Since in the following design examples, reference will be made to unreinforced jet-grouted columns, the stability of the canopy will be considered possible only as long as full overlapping is ensured in the canopy's cross-section and each arch is able to sustain soil loads.

The design goal is to assign the diameter D of the columns, the spacing s among columns, the canopy opening angle β and the span length to obtain a jet-grouted structure able to sustain the soil pressure during excavation for a tunnel of a given diameter D_{tun} placed at a given equivalent depth

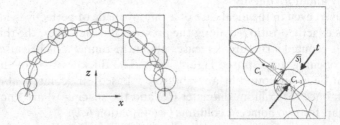

Figure 7.22 Example of the cross-section of a canopy made of 19 jet-grouted columns with defects in diameter and position, and definition of the structural thickness t.

from ground level in a given soil. Once the diameter of the columns has been assigned, the needed number N of columns can be easily determined (in the case of a semicircular tunnel section, $N = 1 + \pi D_{tun}/2s$).

The design can be carried out with a deterministic, semiprobabilistic or probabilistic approach. Whatever the approach, the design goal is to draw jet-grouted canopies with a capacity (defined as the ability of a given structure to carry the imposed loads) larger than the demand (defined as the structural requirements to carry the imposed loads).

With the deterministic approach, partial factors must be introduced to reduce the diameter of columns (Equation 6.5) and the uniaxial compressive strength of the jet-grouted material (Equation 6.10), whereas a design value must be assigned to quantify the deviation of columns from their prescribed direction. Since the combination of these factors may be too conservative with a deterministic approach, semiprobabilistic or probabilistic design approaches may be convenient.

7.3.1.1 Semiprobabilistic approach

This approach takes advantage of the combination of a probabilistic calculation to quantify the geometrical properties of the canopy and of the deterministic solution of a simplified structural scheme. It has been proposed and described in more detail by Costabile et al. (2005), Flora et al. (2007) and Lignola et al. (2008).

As also shown for the other design examples reported in this chapter, once the diameter (in terms of mean value and coefficient of variation) and the initial spacing of columns is assigned, and the shape of the jet-grouted canopy is known (in terms of its initial diameter, of the opening angle β shown in Figure 7.20 and of its statistical parameters), it is possible, with a Monte Carlo procedure, to generate a large number of virtual jet-grouted structures and calculate, in each cross-section arch, at every intersection between adjacent columns, the structural thickness t. For a number of columns n, in each generated structure and in each cross-section of the canopy, n-1 values of t are calculated; with this semiprobabilistic approach, only the minimum of such values (t_{min}) is considered and conservatively considered as the reference thickness of the arch. After all the canopies with defects have been randomly generated, a corresponding number of values of t_{min} is available for each cross-section. Then, assuming a value of the probability of having smaller values (from now on, the value 5% will be considered, but any alternative choice may be done), the minimum thickness $t_{min,5\%}$ related to such a probability and calculated with the Monte Carlo procedure is known for each section. This step represents the probabilistic part of this calculation.

Because the arches get larger along each single canopy, correspondingly decreasing values of $t_{min,5\%}(y)$ are obtained for increasing distance (y) from the initial section, eventually becoming nil if there is a lack of overlapping. In general, the $t_{min,5\%}(y)$ curve depends on the diameter D of columns and on the mutual spacing s between columns, the former having a statistical distribution dependent on the injection parameters and native soil, and the latter influenced by the spacing given in the initial section, the frustum of cone opening angle β and the deviation from the prescribed direction.

The $t_{min,5\%}(y)$ curve can then be assumed to represent the structural capacity of the canopy and, therefore, has to be compared with the demand, which depends on the applied loads, on the geometrical characteristics of the arch, on its base restraints and on the mechanical properties of the jet-grouted body.

The demand can be conveniently represented by a minimum structural thickness t_{nec} needed for equilibrium in any cross section of the canopy and must be found performing a structural analysis. Obviously, t_{nec} increases along the tunnel axis y, as D_{tun} increases with y.

The boundary restraints at the base of the arch are usually frictional: they can be considered as hinges if the horizontal load generates a shear stress τ at the soil to the jet grouting base interface lower than the shear strength τ_{max} (see Figure 7.23); if the limit value $\tau = \tau_{max}$ is attained, horizontal displacements may occur (i.e., the supports behave as rollers) and base instability with structural collapse is considered to occur. As a matter of fact (Flora et al. 2007), this condition is never critical for the typical stress states acting on jet-grouted canopies, and therefore the mechanics of static instability caused by shear failure at the base restraints will not be considered in the following.

The arch with two hinges is a hyperstatic structure one time statically undetermined. A simple procedure to solve this static problem, using the lower bound theorem of plastic analysis and the concept of trust line, is described in detail in Lignola et al. (2008), to which the reader is addressed for more details. This procedure is based on the use of the thrust line (TL) (introduced by Hooke in 1695; Lancaster 2005), defined as the position in an arch of the resultant forces. Because of its simple geometrical meaning, the use of TLs

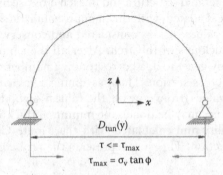

Figure 7.23 Structural scheme of the cross-section of the canopy.

in arch design dates back to the old times of civil engineering. Typically, in masonry arch design, it was possible to overtake the hyperstatic indeterminacy, modelling TLs first and then placing structural material around them, keeping the thrust in the centre of the structures. If nonreinforced jet-grouted canopies are represented by a number of arches, it may be convenient to consider a null tensile strength, and therefore, the simple TL approach is well suited.

In this case, the loads produced by the soil on the arch are considered constant, and the ratio between the horizontal and the vertical effective normal stress components is taken equal to k. The quantification of k is needed to fully define the loading condition. Lignola et al. (2008) have shown that the adoption of the lower boundary theorem (LBT) of limit analysis is conservative and appropriate only if, for every case to be considered, two analyses are carried out, separately considering the limit cases $k = k_a$ (active coefficient of earth pressure) and $k = k_p$ (passive coefficient of earth pressure), that is, considering the minimum and the maximum possible values of k. This is because the LBT can be considered valid only for proportional loading paths, that is, for stress paths in which vertical and horizontal stresses are changed keeping a constant ratio (Chen 1975). Because of soil–structure interaction (SSI) and consequential stress changes around the jet-grouted canopy, the proportionality does not strictly hold in reality, and therefore the use of LBT is conservative only if the two extreme load conditions ($k = k_a$ and $k = k_p$, both representing proportional loading paths) are analysed.

By using k_a to define the horizontal soil stress components, the lowest confinement is given to the structure, and the lowest possible normal stresses are calculated in the structure. On the contrary, by using k_p, the largest axial loads are imposed to the structure. As far as eccentricities are concerned, both limit conditions may be critical (Lignola et al. 2008). The two limit analyses allow analysis of the most critical structural conditions in terms of both eccentricities and axial loads. The minimum necessary structural thickness t will be the largest of the values inferred from the two independent analyses.

Although the aforementioned considerations ensure the conservativeness of the approach, they do not allow the evaluation of how conservative the approach is. This will depend on the values of k, k_a and k_p, as well as on the effects of SSI.

The solution of the static problem gives the demand curve $t_{nec}(y)$. Because the cross-section of the canopy increases along the axis, so does the needed structural thickness $t_{nec}(y)$; as previously shown, on the contrary, $t_{min,5\%}(y)$ decreases along the span. Therefore, three possible design situations may occur, as sketched in the capacity $(t_{min,5\%})$–demand (t_{nec}) plot shown in Figure 7.24:

1. At some distance from the initial section of the canopy, the structure is not fully overlapped, and the arch is not continuous to the end of the span. Furthermore, $t_{min,5\%} \geq t_{nec}$ only to point A in Figure 7.24.

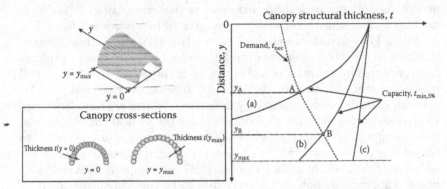

Figure 7.24 Scheme of arch structural demand t_{nec} versus capacity $t_{min,5\%}$ curves indicating three different design situations: (a) Arches not fully continuous to the end of the span and capacity larger than demand only to a distance y_A (maximum allowable canopy length); columns geometry to be refined. (b) Arches fully continuous to the end of the span but capacity larger than demand only to a distance y_B (maximum allowable canopy length); columns geometry possibly satisfactory with a shorter span. (c) Arches fully continuous and capacity larger than demand to the end of the span (optimum design).

Then, with the static scheme considered, point A defines the maximum distance y_A from the origin at which the arch is in equilibrium with the soil loads, that is, the maximum operational canopy span for the considered set of columns and statistical parameters. If the designer wants to use the diameter and spacing (and, therefore, the number of columns) adopted in this case, then there is no reason to make the columns longer than $y_A + a$ (a being the overlapping distance between subsequent canopies). As a consequence, canopies shorter than y_{max} should be created. From a constructional point of view, if the distance between the starting section of two subsequent canopies is reduced, a correction must be introduced by increasing the opening angle β (see Figure 7.24). Since an increase of β will certainly result in values of $t_{min}(y)$ smaller than the ones previously calculated, there is the need to recalculate the capacity curve, and so an iteration in the design process is necessary.

2. The structure is continuous to y_{max} (end of the canopy) but is able to sustain the loads only to point B which, therefore, defines the maximum operational span. Again, there is no reason to make columns longer than $y_B + a$; the span between canopies could then be modified with the same considerations made for the previous case.

3. The structure is continuous to y_{max} and is always able to sustain the loads ($t_{min} \geq t_{req}$ for $0 \leq y \leq y_{max}$). The set of columns is well suited for such a desired span.

Cases 1 and 2 can be considered as effective solutions, therefore ending the canopy design process, only if the resulting length of the span is not too short. However, since in most cases it would not be convenient for the construction sequence to have very short spans, it is usually the case to fix the desired value of y_{max} as a design constraint. Then, to be neither over-conservative nor unsafe in design, the optimum design goal, in this case, should be that of imposing equilibrium in the most critical cross-section (at $y = y_{max}$), that is, imposing $t_{min,5\%}(y_{max}) = t_{nec}(y_{max})$.

Because $t_{nec}(y)$ depends on the applied loads, on the arch geometry and on the mechanical properties of the jet-grouted material, once these are given, the necessary thickness cannot change. Then, the design issue of having resistant arches to the end of the canopy (and not using reinforced columns, which would change the static scheme) can be met by acting on the capacity curve $t_{min,5\%}$ by one or more of the following options: assigning a larger column diameter, reducing the spacing among columns and/or increasing the thickness of the shell by means of a multilayered canopy.

Whereas the first two options are rather obvious and need no further comments, just leading to a new calculation of $t_{min,5\%}(y)$, the latter option can be accomplished by modifying the design geometry either by injecting concentric rows of columns from the same tunnel section (Figure 7.25a) or

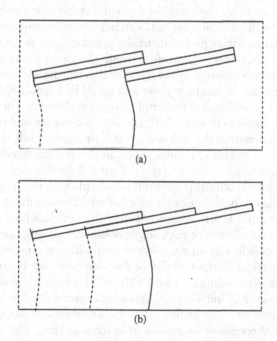

(a)

(b)

Figure 7.25 Multilayered canopies formed by creation of multiple rows of columns from the same tunnel section (a) and overlapping of subsequent canopies (b).

by increasing the overlapping between subsequent canopies at least up to half of their total length (Figure 7.25b). In both cases, the supporting arch is made by two concentric rows of columns. However, the opening angle β and the deviation of columns from their prescribed directions should be carefully considered because a very small overlapping between adjacent columns may occur from the combination of these factors.

For the sake of simplicity, the total thickness of the canopy $t_{tot}(y)$ can be considered equal to the sum of the single values pertaining to each layer of columns or may be computed with the previously described Monte Carlo procedure, considering the overall geometrical complexity.

7.3.1.2 Probabilistic approach

With this approach, the geometry of the canopy is generated with the usual Monte Carlo procedure. Each of the large number of generated structures, having an irregular shape and variable thickness, is then divided in subsequent arches, obtained as cross-sections of such jet-grouted structures. Then, each of these irregular arches, for each of the generated structures, is statically analysed. The internal stress state (defined by a combination of axial thrust and bending moment) is calculated in each element of any single arch, and the most critical structural situation along any arch axis, in terms of eccentricity (bending) and normal stress (compression), is considered. Once such a calculation has been performed in any cross-section of all the irregular canopies probabilistically generated, then the most critical internal stress state at any position y along the canopy's axis for a given probability level is known, and the structural performance can be assessed. This procedure, which has been shown in detail by Lignola et al. (2009), is fully probabilistic and makes no simplifying hypotheses on the shape of the canopy, thus leading to more realistic results. As a matter of fact, the irregular shape induces within the arches a stress state significantly different from the one related to an ideally regular shape, and therefore, bending moments and stress eccentricities may be larger. Figure 7.26 shows the mechanical idealisation adopted with this method, which analyses the structural problem considering the behaviour of a jet-grouted structure made of a number of elements of finite length connected over the nodes, which are assumed to be located at the centre of each single column. The supporting structure is, therefore, modelled as an assembly of monodimensional beam elements (for the geometrical characteristics of the elements, see Figure 7.26) that are able to sustain bending moments (M), axial forces (N) and shear forces (T). The behaviour of individual elements is characterised by the element's stiffness relationship. The analytical and computational developments are best executed throughout by means of matrix algebra. The displacement or stiffness method is, by far, the most popular one because of its ease

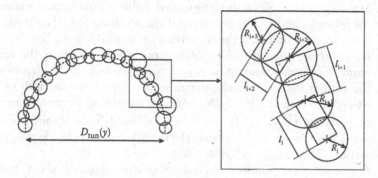

Figure 7.26 Structural schematisation of the irregular cross-section arch of the canopy at the generic abscissa y. The elements are joined in the centres of the columns, each having a length *l* equal to the spacing *s* between the columns and a thickness equal to the overlapping thickness *t*.

of implementation as well as of its formulation for advanced applications (Lignola et al. 2009).

As with the semiprobabilistic approach, at the base of each arch frictional restraints can be considered, with a limit shear stress value τ_{max} given by the Mohr–Coulomb criterion. As a consequence, the restraints behave as hinges as long as $\tau < \tau_{max}$.

The stress field is assumed to be constant, and therefore the vertical q_v and horizontal q_h effective stresses are

$$q_v(z) = \gamma \cdot (H - z)$$
$$q_h(z) = k \cdot (H - z) \tag{7.7}$$

As is well known and previously recalled, the stress state around the tunnel certainly differs from this. However, in Equation 7.7, H should be considered more as an equivalent rather than as a true tunnel depth. Furthermore, since with this approach it is relatively easy to consider SSI, the initial value of k can be taken as k_0, allowing its variation caused by the structure deformation.

The safety check of the jet-grouted structure can be carried out considering the following four mechanisms:

1. *No overlapping*: In the two-dimensional arch scheme, the first obvious reason of failure is the lack of continuity, that is no overlapping at least between two adjacent columns ($t_i = 0$ in any position of any of the cross-section arches).

2. *Stability failure*: Since the jet-grouted soil is a non-tension material, the overall stability is not granted if the resultant force is outside the cross-section, that is, if the eccentricity is $e_i = |M_i|/N_i > t_i/2$.

3. *Jet-grouted material failure*: Structural failure occurs if the maximum calculated compressive strength σ_{max} in one of the elements of the arch is higher than the unconfined compressive strength of the jet-grouted soil. If $e_i = |M_i|/N_i < t_i/6$, then cracking cannot occur and $\sigma_{max,i} = N_i/t_i + 6|M_i|/t_i^2$. If cracking is allowed, then stability is still granted if $t_i/6 < e < t_i/2$ because the resultant force is inside the cross-section. Then $\sigma_{max,i} = 4N_i/(3t_i - 6e_i)$.

4. *Frictional base failure*: According to the classical shear theory (Timoshenko and Goodier 1987), the shear stress can be evaluated in the base rectangular section as $\tau = 1.5|H|/t$, where H is the horizontal base reaction force. If frictional failure is attained at the base restraint ($\tau = \tau_{max}$), then the arch is supposed to be unstable.

7.3.1.3 SSI in the probabilistic approach

The interaction between the canopy and the surrounding soil may be considered through independent Winkler-type connecting springs. Then, the stresses acting on the structure change along the loading process because of structure's deformation. Inward movements release the springs and decrease the stress (with a lower bound given by the active state), whereas outward movements result into an increase (with an upper bound given by the passive state). The SSI may influence the stress state in the structure to a great extent, its relevance depending on soil–structure relative stiffness.

The deformed shape $w(\eta)$ in the local coordinate system of an element loaded by a transverse unit displacement is given by

$$w(\eta) = 2x^3/L^3 - 3x^2/L^2 + 1 \tag{7.8}$$

Because of this displacement field, the independent springs load the element with a pressure $p(\eta) = \kappa \cdot w(\eta)$, where κ is the subgrade stiffness, giving rise to an additional displacement field, $w'(\eta)$, according to the equation

$$w'(\eta) = \frac{\kappa L^4}{EI}\left(\frac{x^7}{420L^7} - \frac{x^6}{120L^6} + \frac{x^4}{24L^4} - \frac{13x^3}{210L^3} + \frac{11x^2}{420L^2}\right) \tag{7.9}$$

In principle, the subgrade stiffness is a nonlinear function and depends on the sign of the displacement function. Different spring models are available to model the interaction. In this case, to avoid an iterative approach, the simplest case of a constant Winkler spring stiffness up to the limit stress condition (either active or passive) is considered. The spring stiffness κ can be considered to depend on the single beam element width B and on soil stiffness E_{soil} via the expression

$$\kappa = 2 \cdot B \cdot (E_{soil}/D_{tun}) \tag{7.10}$$

It can be shown (Lignola et al. 2008) that the additional pressure $p'(\eta) = \kappa \cdot w'(\eta)$ is negligible (i.e., smaller than $p(\eta)/100$) if $(\kappa \cdot L^4) > (7 \cdot EI)$. Most times, this inequality is satisfied and, as a consequence, no further additional displacement field, $(w''(\eta))$, needs to be considered to account for SSI.

7.3.1.4 Calculation example with the probabilistic approach

In this example, the simple case of a tunnel having $D_{tun}(y = 0) = 8$ m and $H = 16$ m is considered, with a canopy having a length of 15 m, an overlapping with the adjacent canopies of 3 m and therefore an effective length of 12 m. A total number of 38 columns having $D = 0.6$ m is considered as forming the canopy. The opening angle of the canopy (see Figure 7.20) is $\beta = 5.71°$. The structure is studied by dividing such an effective length in 12 arches (one for each metre). The soil has $\gamma = 18.5$ kN/m³ and $\varphi' = 35°$. The probabilistic approach just described has been used, assuming CV(D) = 0.10, SD(β) = 0.5.

Figure 7.27 reports the deformed shapes of the structure with and without SSI for the canopy with and without defects (the latter being nonsymmetric). As expected, SSI reduces, to a great extent, the displacement field.

It is worth pointing out that the reduced displacement field yields to less critical arch structural conditions, and so it may be worth to take SSI into account to avoid excessive conservativeness.

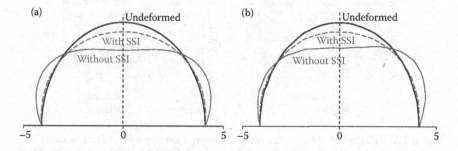

Figure 7.27 Deformed shape of the tunnel for the calculation example, showing the effect of soil–structure interaction (SSI). (a) Structure with no defects; (b) structure with defects; results for a level of confidence of 95% (tolerated probability of failure of 5%).

The results of each Monte Carlo individual simulation can be aggregated into the probability of failure P_f of each of the four mechanisms previously described.

In the case of an ideal geometry of the structure (no defects), for the considered example, no failure is detected along the entire canopy. Figure 7.28 reports the results of the probabilistic analysis in a synthetic way. Each of

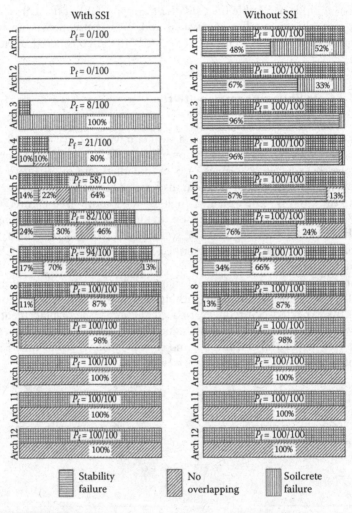

Figure 7.28 Probability of failure for arches forming a canopy with defects in sandy soil with λ = 2, CV(D) = 0.10 and σ(ψ,ξ) = 0.5. Effect of SSI and details of the different failure mechanisms along the canopy axis. (Modified from Lignola, G. P. et al., Effects of defects on structural safety of jet-grouted umbrellas in tunelling. *Proceedings of the 10th International Conference on Structural Safety and Reliability*, Osaka, Japan, Paper CH454: 13–17 September, 2009.)

the 12 coupled bars reported in the figure refers to the arches (width, 1 m) in which the structure is divided, arch 1 being the one located at the beginning of the analysed canopy ($y = 0$). The upper bar represents the overall probability of arch failure P_f for the given cross section/arch of the canopy, whereas the lower bar represents the percentage of such a probability linked to one of the three expected failure modes (jet-grouted material failure, stability failure, no overlapping), the frictional failure mechanism at the arch bases being never triggered.

The figure clearly shows that, in this case, the large scatter in column position results into a structure that has holes (i.e., no structural continuity and, therefore, a missing arch), starting from the seventh arch ($y = 7$ m). If no SSI is considered, for the considered example, the situation is critical from the very beginning of the canopy because the probability of failure is always 100%. Considering SSI, the results are just a little less critical, the probability of failure being too large to be accepted from the fourth arch ($y = 4$ m).

A lower spacing among columns, or better a larger diameter of columns should then be assigned to have longer operational spans.

7.3.2 Jet-grouted shafts

The working principle of jet-grouted shafts is not dissimilar from that of tunnel canopies, apart from the direction of columns: a cylindrical vertical shell of jet-grouted material is, in fact, created by a sequence of partially overlapped vertical columns arranged on a circular plan to temporarily support the soil during the excavation of the shaft (Figure 7.29a). While performing excavation, the supporting function of the jet-grouted structure may be enhanced by the insertion of steel ribs placed on the inner surface of the shaft. Sometimes, the flexural resistance of columns is improved by inserting steel reinforcements (see Chapter 6.4.4).

Apart from these additional countermeasures, the earth-retaining capability of the jet-grouted structure lies on the formation of a continuous cylindrical shell, able to sustain the external loads. Thus, structural continuity, that is, the overlapping among adjacent columns at all depths is a fundamental requirement.

The jet-grouted structure can be simply modelled by superimposing longitudinal and cross-sectional elements, the latter represented by the rings made of partially overlapped columns (see Figure 7.29a), as for tunnel canopies (Section 7.3.1). The surrounding earth pressure is sustained by the combination of these two groups of mutually interacting elements, each supporting a fraction of the total load. The interaction depends on load variation with depth, on relative stiffness of the different elements and on its variation with depth. The interaction between radial and longitudinal elements is represented by the scheme reported in Figure 7.29b; for the ideal

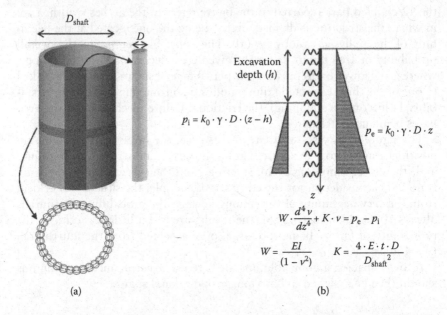

Figure 7.29 Structural idealisation of a jet-grouted shaft made of superimposed arches and longitudinal elements (a) and interaction between longitudinal and cross-sectional elements (b).

case of a perfectly cylindrical shaft and a linear stress–strain response of the jet-grouted material, the radial response of the horizontal rings may be represented by linear springs whose reactive stiffness (k) is dependent on the diameter of the shaft (D_{shaft}), the thickness created at the intersection of columns (t) and the stiffness of the jet-grouted material (E); the longitudinal element can be reasonably considered as a beam subjected to the earth pressure and resting on the previously defined springs (Winkler-type model).

Figure 7.30 reports the results obtained with this simple structural scheme for the particular case of D_{shaft} = 12 m, D = 0.6 m, t = 0.50 m (corresponding to a spacing between columns s = 0.45 m), E_{jg} = 2500 MPa, ν_{jg} = 0.2, γ = 20 kN/m³ and k_0 = 0.5.

The results have been plotted considering different excavation depths into the shaft (5, 10, 15 and 20 m from ground surface). In the case of no defects along the shaft (Figure 7.30a), bending moments are very low. However, if overlapping among columns is lacking at some depth (this effect has been artificially introduced in the calculation by assigning a nil value to the spring's stiffness at the depth of 15 m), a significant increase of the bending moment will locally occur on the columns (Figure 7.30b),

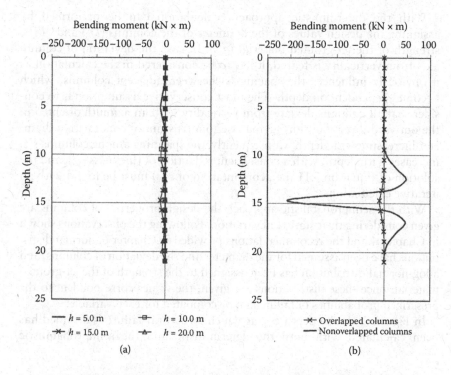

Figure 7.30 Longitudinal bending moment in columns computed with the scheme of Figure 7.29b (D_{shaft} = 12 m, D = 0.6 m, t = 0.51 m, E = 2500 MPa and ν = 0.2) for the case of completely overlapped columns (a) and a lack of overlapping at the depth of 15 m (b).

which may become critical if the ultimate flexural moment of the columns is exceeded.

Figure 7.30 clearly shows on the contrary that, in the case of no holes in the jet-grouted shaft, bending moments are negligible. It is therefore reasonable to study the behaviour of the structure by looking to its cross-section, that is, to a circular beam of thickness t, symmetrically loaded by the horizontal earth pressure. Such a jet-grouted structure is loaded only in compression, and structural failure may be attained only if the limit compressive strength of the jet-grouted material is reached. In the ideal condition of a perfect cylinder of constant thickness t, considering an earth pressure linearly increasing with depth z, failure of the jet-grouted material is attained at a depth z_{lim} expressed as follows (Croce et al. 2006):

$$z_{lim} = \frac{2q_u\sqrt{D^2 - s^2}}{k_0 \cdot \gamma \cdot D_{shaft}} \tag{7.11}$$

With the deterministic approach, calculation can be performed by assigning the design values of the diameter of the columns (D_d) and of the strength of the jet-grouted material (q_{ud}). In case the deviation of column axis from verticality is feared, it has to be considered in the calculation as it obviously influences the spacing (s) between adjacent columns, which becomes dependent on depth. The most conservative assumption is to consider that all columns deviate from verticality with an azimuth oriented in the outward direction, thus giving rise to a frustum of cone, with a diameter increasing with depth. Consequently, the spanning among columns $s(z)$ increases with depth, with a consequent reduction of thickness $t(z)$, and the solution of Equation 7.11 (i.e., computation of z_{lim}) must be found with an iterative calculation.

With the semiprobabilistic approach the design properties of columns are given considering their statistical variation. Following the observations shown in Chapter 4 and the recommendations provided in Chapter 6, normal distributions have been assumed for the diameter and the deviation of columns, and a log-normal distribution has been assigned to the strength of the jet-grouted material; once these distributions are given, the values correspondent to the prescribed probabilities of failure can be computed for each variable.

In Figure 7.31, the depth z_{lim} at which structural failure is attained has been calculated with both the deterministic and the semiprobabilistic

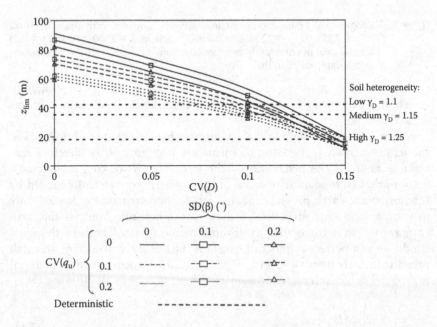

Figure 7.31 Limit length of a jet-grouted shaft (D_{shaft} = 12 m, D = 0.6 m, s_0 = 0.44 m, q_u = 13.4 MPa).

approach, considering $D_{shaft} = 12$ m, $k_0 = 0.5$, $\gamma = 20$ kN/m³, span at ground level $s_0 = 0.44$ m, and assigning the characteristic values of the column diameter and the uniaxial compressive strength, respectively, equal to $D = 0.6$ m and $q_u = 13.4$ MPa.

In the deterministic calculation, the diameter of the columns has been scaled by the partial factors suggested in Table 6.5 ($\gamma_D = 1.10$, 1.15 and 1.25 for soils with, respectively, low, medium and high degrees of heterogeneity and poor experimental investigations). A systematic outer divergence with a deviation angle $\beta = 0.4°$ has been assigned to all columns, and the uniaxial compressive strength of the jet-grouted material q_u has been scaled by the partial coefficient $\gamma_M = 1.5$ (see Chapter 6.4.3).

In the semiprobabilistic calculation, the values of D and q_u have been computed as those corresponding to the 5% fractile of their distributions, whereas the value corresponding to the 95% fractile of its distribution has been taken for β. The above given characteristic values ($D = 0$, 6 m; $q_u = 13.4$ MPa, $\beta = 0$) have been taken as mean values of the distributions, and calculation has been performed for different standard deviations [CV(D), CV(q_u) and SD(β)].

The smooth decay of the semiprobabilistic curves shows that the variation of diameter is more relevant than the deviation of column axes and the strength of the jet-grouted material. In the deterministic approach, these defects are accounted for by the partial factors γ_D and γ_M producing a significant reduction of z_{lim}. A comparison between the two approaches (deterministic and semiprobabilistic), which can be done by considering the coefficients of variation typically observed on the diameter of columns for materials having different degrees of heterogeneity (see Table 4.9), shows that the deterministic approach gives lower limit lengths and is, thus, generally more conservative.

More complex calculation has been implemented by Croce et al. (2006), who performed a probabilistic analysis of a similar structure. The authors performed their analysis by randomly generating structures with the Monte Carlo method, by computing the stress distributions at different depths with a finite element code and by checking the continuity and stress compatibility of the jet-grouted material. Their results show a meaningful role of the deviation of the cross-sectional rings from perfect circularity, which turns into higher compressive and/or tensile stresses.

7.3.3 Three-dimensional effects

The previously shown two-dimensional analyses of nonreinforced curved jet-grouted retaining structures (based on the simplified schemes reported in Figures 7.21 and 7.30) assume that a lack of continuity, even in a single point, leads to structural collapse in that section. Indeed, this may be too conservative because the three-dimensional jet-grouted tunnel canopies or vertical shafts may carry soil loads even if they are not continuous in small portions, considering their three-dimensional behaviour. The main questions are how small should this portion be? How can these discontinuities be considered from a structural viewpoint?

In principle, the answer can be given only by performing fully three-dimensional analyses, reproducing in the most realistic manner the sequence of operations (making of columns, excavation, insertion of additional reinforcements), the geometrical and mechanical variability of the jet-grouted structures and the three-dimensional response of the surrounding soil. This seems however too complex for practical design.

A simplified analysis has been performed by Flora et al. (2012b), who adopted the analytical method suggested by Young and Budynas (2002) to analyse the maximum compressive stresses generated around a circular hole in a jet-grouted structure. With this study, it was found that, unless the jet-grouted material has a particularly low strength, holes of dimensions of a metre or so do not necessarily imply structural failure, even at considerable depths. However, this result cannot be generalised, and the two-dimensional analyses remain the only ones that can be easily carried out in current practice. The results of the three-dimensional analyses indicate that there is some extra safety factor hidden in the assumption of a two-dimensional scheme.

7.4 HYDRAULIC CUTOFFS

As pointed out in Section 5.4, jet-grouted cutoffs can be created by joining columns or panels, combined according to various possible shapes (linear, wedge, cellular, etc.). However, the most frequent type of jet-grouted cutoff is made of one or more parallel rows of partially overlapped cylindrical columns.

It is recalled that the jet-grouted material can be considered as being practically impervious, with the possible exception of treatments performed in clean gravels, whose soil pores may not be completely filled by grout (see Chapters 3 and 4). However, the continuity of jet-grouted cutoffs may be compromised by the defective overlapping of adjacent columns, which may be caused by local diameter reductions and/or deviations of the column axes from verticality. The design of jet-grouted cutoffs should thus account for these physiological defects of the jet columns to guarantee the continuity of the seepage barrier. This task can be accomplished by performing a specific analysis of the barrier continuity by means of deterministic, semiprobabilistic or probabilistic approaches, as described in the following sections.

7.4.1 Single row cutoff: Deterministic approach

The deterministic analysis of the cutoff continuity can be accomplished by combining the previously reported Equations 6.1, 6.2 and 6.4. For a single row cutoff, the analysis can be based on the most critical overlapping conditions for a couple of adjacent columns. In this simple case, the minimum barrier thickness t is obtained when two contiguous columns diverge to opposite azimuth directions ($\alpha_1 - \alpha_2 = 180°$ in Equation 6.4):

$$t(z) = \sqrt{D^2 - \left[s_0 + 2 \cdot z \cdot (\tan\beta_1 + \tan\beta_2) \right]} \tag{7.12}$$

This equation has been used to calculate the normalised thickness t/D for typical values of diameter D, spacing s_0 (taken at ground level, i.e., $z = 0$) and columns having the same deviation ($\beta_1 = \beta_2 = \beta$). The results are reported in Figure 7.32, showing the relevance of the column axes deviation, whose influence increases with the depth of the cutoff. For the largest assumed inclination angles ($\beta = 0.5°-1°$), the panels are unable to prevent seepage even for limited depth ratios ($z/D < 20$). These plots may be used for a preliminary design of the jet-grouted cutoff once design values of the diameter (D_D) and of the inclination of columns (β_D) have been assigned.

With the deterministic approach, the diameter D reported in the plot of Figure 7.32 may be taken as the design value D_d (see Chapter 6.4.1), obtained by scaling the characteristic value D_k by a partial factor γ_D (see Table 6.5). The design inclination β_D can be elected following the indications given in Section 6.4.2.1.

7.4.2 Single row cutoff: Semiprobabilistic approach

The semiprobabilistic approach can be adopted if the statistical distributions of column diameters and inclinations are known. The calculation is performed by selecting design values D_d and β_d, corresponding to the given probabilities of failure (i.e., to arbitrarily assigned fractiles of the assumed distributions).

An example is reported in the following, assuming as design values of D and β those corresponding to a 5% fractile of their distributions. The normal probability function has been given to model the statistical distribution of column diameters (with a mean value D_{mean} and a coefficient of variation CV(D)) and of the deviation angles (with a mean value β_{mean} and a variable standard deviation SD(β)), whereas a uniform distribution has been assigned to the azimuth angle α.

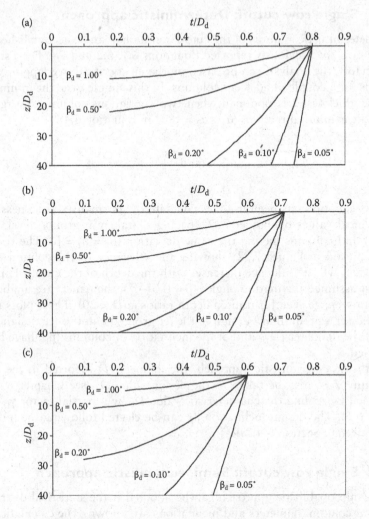

Figure 7.32 Deterministic approach: normalised thickness *t/D* of jet grouting panels formed by a single row of columns (a) $s_0/D = 0.6$; (b) $s_0/D = 0.7$; (c) $s_0/D = 0.8$.

The role of the different factors (D_{mean}, CV(D) and SD(β))can be explicitly considered by expressing the thickness $t(z)$ of a cutoff formed by a single row of columns as follows:

$$t(z) = \sqrt{D_{\mathrm{mean}}^2 [1 - 1.65 \cdot \mathrm{CV}(D)]^2 - \left(s_0 - z \cdot \sqrt{2 \cdot [1 - \cos(0.95 \cdot \pi)]} \cdot \tan^2\left(1.65 \cdot \mathrm{SD}(\beta)\right)\right)^2}$$

$$(7.13)$$

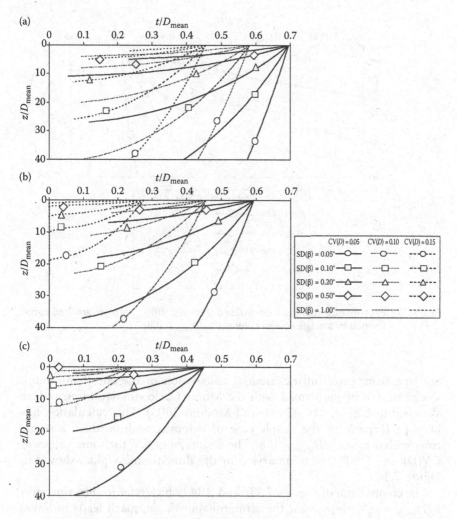

Figure 7.33 Semiprobabilistic approach: normalised thickness t/D_{mean} of jet grouted cut-offs formed by a single row of columns, (a) s_0/D_{mean} = 0.6; (b) s_0/D_{mean} = 0.7; (c) s_0/D_{mean} = 0.8.

The plots reported in Figure 7.33 show the combined effect of the uncertainties on the diameter and the inclination of the columns.

7.4.3 Single row cutoff: Probabilistic approach

Another possible way for assessing cutoff continuity is to adopt the fully probabilistic approach, as previously shown for other jet-grouted structures.

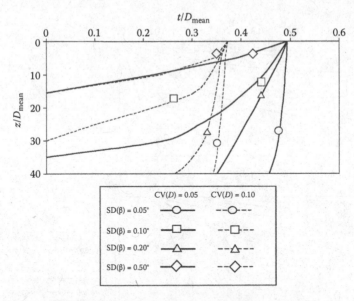

Figure 7.34 Probabilistic approach: normalised thickness t/D_{mean} of a jet-grouted cutoff formed by a single row of columns (s_0/D_{mean} = 0.8).

Starting from probabilistic models, calculation of the cutoff minimum thickness can be performed with the Monte Carlo simulation procedure sketched in Figure 6.10 (Croce and Modoni 2005). This calculation has been performed for the simple case of columns positioned on a single row, with a ratio s_0/D_{mean} = 0.8. The results obtained for some values of CV(D) and DS(β) are summarised in the dimensionless plots shown in Figure 7.34.

The comparison of Figures 7.33c and 7.34 (which refer to the same ratio s_0/D_{mean} = 0.8) shows that the semiprobabilistic approach leads to lower values of the cutoff thickness t, and it is thus more conservative than the probabilistic approach. In addition, the latter approach provides a more accurate analysis, and it may thus be concluded that the fully probabilistic approach grants the most realistic and yet most convenient analysis of cutoff continuity. A direct comparison with the deterministic approach is not easy because the result of the latter depends on the chosen values of the partial factor γ_D and of the deviation angle β_D.

7.4.4 Double row cutoff

If a single row cutoff is considered too risky, a possible alternative solution consists of placing a second row of columns adjacent to the former one. The two parallel rows are usually shifted, as sketched in Figure 7.35, creating a triangular array characterised by higher filling ratios compared with the rectangular array (see Figure 6.4). Assuming that the water can find its way only where there is no overlapping, effectiveness is improved because the path that the water must follow to cross the barrier is much longer than in the case of a single row. This seepage path can be considered as the equivalent thickness of the cutoff. In particular, in the case of ideally cylindrical vertical columns of the same diameter forming an equilateral triangular cell, the equivalent thickness is $t_{tot} \approx 3t$, t being the thickness of the single row cutoff.

The probabilistic approach can be performed by assigning a fixed spacing value s_0 at ground level and by assuming appropriate probabilistic distributions for the diameter D, the azimuth α and the axis deviation angle β, as suggested in Sections 6.4.1 and 6.4.2. Cutoff failure takes place where a flow path is found at some depth z ($t_{tot}(z) = 0$).

The results obtained by generating 1000 geometrical configurations with the Monte Carlo routine are reported in Figure 7.36. These analyses have been carried out with reference to the example of a cutoff made of two parallel rows with a triangular cell, each formed by 20 overlapped columns spanned at ground level $s_0/D_{mean} = 0.8$ (Figure 7.36a). The statistical parameters adopted in the example are reported in the Figure 7.36b (the azimuth angle α needs no parameter being uniformly distributed).

The figure confirms that the most relevant factor affecting cutoff continuity is column inclination, thus suggesting that this is the key parameter to be controlled on site. In this example it is, in fact, practically impossible to produce impervious cutoffs, even of limited depths, if the standard deviation SD(β) is larger than 0.5°. With such large deviations, the scattering of diameters is not of particular concern, whereas it becomes progressively

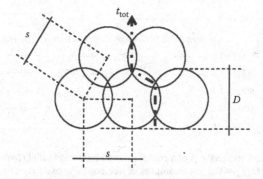

Figure 7.35 Scheme of a cutoff formed by a double row of columns and indication of the possible seepage path (dotted line).

more important if the deviation from verticality becomes negligible. For instance, for CV(D) equal or larger than 0.20, the probability of cutoff discontinuity is larger than 6%, irrespective of the values of SD(β). Figure 7.36b can be considered as a design chart in which, after fixing an acceptable probability of discontinuity, the maximum allowable depth to have water tightness can be found for a given set of statistical values of the random variables D and β.

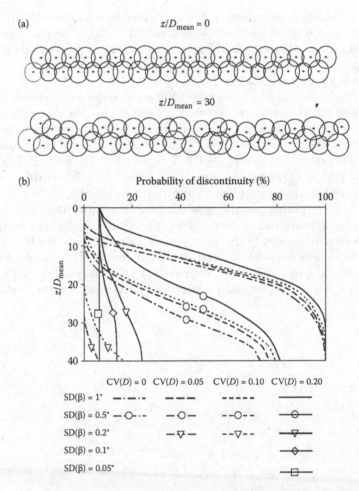

Figure 7.36 Probabilistic analysis of a cutoff formed by two parallel rows of columns having s_0/D_{mean} = 0.8. (a) Examples of horizontal cross-sections at two different depths for CV(D) = 0.1, SD(β) = 0.2°; (b) probability of cutoff discontinuity.

7.5 BOTTOM PLUGS

7.5.1 Mechanical scheme and design goals

When planning an open excavation in granular soils below groundwater level, one of the main design concerns refers to water seepage from the bottom. Such a seepage may cause local piping, subsidence of the surrounding area with undesired settlements of nearby buildings, or a water inflow that may interfere with constructional operations.

The jet-grouted bottom plug created to prevent these effects must be designed to be impermeable and to resist the uplift water pressure. During excavation, this barrier has the further positive effect of bracing the retaining structure, thus carrying out also a static function.

A number of alternative design schemes can be adopted for creating, bottom plugs, as already reported in Chapter 5 (see Figure 5.16). The sketch of Figure 7.37 considers the example of an excavation to be carried out to a depth h_{exc} from ground level, h_1 being the excavation depth with respect to the groundwater table. The jet-grouted bottom plug has a thickness h_3 and is provided with uplift anchors. A layer of untreated soil of thickness $(h_2 - h_3)$ is also resting on top of the plug.

The bottom plug must be able to balance the uplift resulting force caused by the pore-water pressure. Neglecting the contribution of the shear stresses

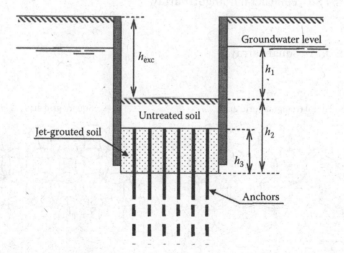

Figure 7.37 Schematic drawing of a jet-grouted bottom plug provided with anchors, considering the possible existence of anchors and of untreated soil on top of the jet-grouted material.

acting at the interface between the bottom plug and the retaining structure, the uplift safety factor SF_{up} is

$$SF_{up} = \frac{\gamma_{jg}h_3 + \gamma_s(h_2 - h_3) + \dfrac{\sum F_i}{A}}{\gamma_w(h_1 + h_2)} \qquad (7.14)$$

in which γ_s, γ_{jg} and γ_w are the unit weights of, respectively, soil, jet-grouted material and water; F_i is the uplift resistance of each anchor; A is the area of the bottom plug. In the simplest case of a completely treated plug ($h_2 = h_3$) with no anchors, Equation 7.14 becomes

$$SF_{up} = \frac{\gamma_{jg}}{\gamma_w} \frac{h_2}{(h_1 + h_2)} \qquad (7.15)$$

Clearly, for the sake of equilibrium, it should be $FS_{up} > 1$. However, higher values of the safety factor may be required by various codes of practice.

To have a massive jet-grouted element of thickness h_2 (Figure 7.38), overlapping of adjacent columns has to be ensured to the depth ($h_{exc} + h_2$). Then, the critical design issue is the choice of a column's spacing s and diameter D suited to this aim, which is obtained by assigning (see Figure 6.4)

$$\frac{s}{D} \le 0.86 \quad \text{equilateral triangular array}$$

$$\frac{s}{D} \le 0.71 \quad \text{square array}$$

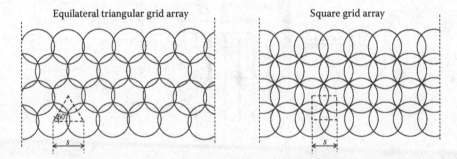

Equilateral triangular grid array　　　　　Square grid array

Figure 7.38 Equilateral triangular and square column centre grid arrays. The schemes have been drawn for the minimum overlapping necessary to have a completely treated cross-section (s/D = 0.87 and s/D = 0.71, respectively, for the triangular and square grids).

The design of a jet-grouted bottom plug may thus appear to be an easy task. In practice, however, the scatter of diameter and orientation of column axis may induce relevant discontinuities that should be considered in design. In particular, the true position of the single columns may differ from the ideal one even at the top of the jet-grouted plug because the columns have to be injected from ground level prior to excavation. As a matter of fact, the bottom plug thickness may thus vary along the excavation area because of typical jet column defects, reaching some minimum thickness $h_2^* < h_2$ (Figure 7.39). As a consequence, the true uplift safety factor SF_{up}^* can be lower than expected and more difficult to calculate.

Assuming that the untreated portions of the plug are interconnected, which is reasonable considering that the divergence of columns from verticality tends to form spaces similar to those depicted in Figure 7.39, the safety factor can be computed, considering the equilibrium of the unit cell (equilateral triangular, or square array in the two schemes of Figure 7.38):

$$SF_{up}^* = \frac{\gamma_{jg}}{\gamma_w} \frac{(A_{cell,0} \cdot h_2 - V_v)}{\left[A_{cell,0} \cdot (h_1 + h_2) - V_v \right]} \tag{7.16}$$

in which $A_{cell,0}$ is the area of the unit cell of the grid at ground level ($z = 0$) and V_v is the volume of the untreated soil that can be found in the cell. Equation 7.16 has been written in the conservative hypothesis that voids are all connected to the aquifer. $A_{cell,0}$ can be easily computed as a function of the spacing among columns (s_0):

$$A_{cell} = s_0^2 \cdot \cos\frac{\pi}{6} \cdot sen\frac{\pi}{6} = \frac{\sqrt{3}}{4} s_0^2 \quad \text{equilateral triangural array}$$

$$A_{cell} = s_0^2 \quad \text{square array}$$

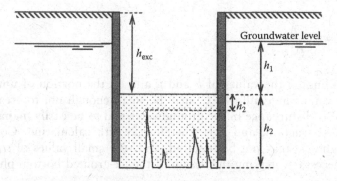

Figure 7.39 Example of a jet-grouted bottom plug with typical defects: nominal (h_2) and minimum (h_2^*) thickness of the impervious barrier.

The calculation of V_v may be cumbersome, having such a volume an irregular shape. Some indications useful to this aim will be given in the following, depending on the adopted calculation choice (i.e., deterministic, semiprobabilistic or probabilistic).

At the extreme, it may happen that there is untreated soil all the way to the top of the plug ($h_2^* = 0$ in Figure 7.39), that is, water can find its way through the plug and the barrier is not impervious any more. This extreme case does not necessarily correspond to a limit condition, depending on the area of the holes (that is, on the true filling ratio F), on the permeability of the untreated soil (k) and on the hydraulic gradient j. Specific calculations have to be carried out to establish if the flow through the untreated soil is tolerable or not and if seepage is able to activate internal piping in the untreated soil (first, in the plug, then, in the surrounding soil). For the unit cell, the water flow Q through the defected plug can be calculated as:

$$Q = k \cdot j \cdot A_{un,top} \tag{7.17}$$

where k is the Darcy permeability coefficient of the untreated soil, j is the hydraulic gradient within the hole in the bottom plug and $A_{un,top}$ is the area of untreated soil in the unit grid at a depth $z = h_{exc}$ (i.e., at the top of the bottom plug), in the reasonable hypothesis that it is the smallest one along the defected plug height.

The hydraulic gradient j through the untreated soil height can be calculated in the simplified (but reasonable and certainly conservative) hypothesis of a nil water pressure head on top of the barrier ($z = h_{exc}$) and of a water pressure head below the barrier equal to that of the surrounding soil, that is, unaffected by the permeation process (see Figure 7.40). With these hypotheses, $j = h_1/h_2$.

The total water inflow into the excavation of area A may be calculated in a simplified manner as

$$Q_{tot} = Q \cdot \frac{A}{A_{cell}} \tag{7.18}$$

Depending on the values of k and j, and on the portion of untreated soil, the total water inflow Q_{tot} may be small enough not to create any measurable disturbance in the groundwater and to be easily manageable during construction, and it is, therefore, worth calculating. Complete water tightness may not be required, and very small values of s/D may not be necessary, making the design of the jet-grouted bottom plug cost effective.

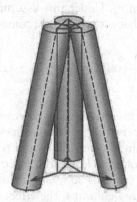

Figure 7.40 Conservative design scheme to be considered for the deterministic design of a bottom plug having a triangular grid: the columns have an assigned inclination to the vertical β_d, the azimuth being oriented in the most critical direction.

7.5.2 Deterministic approach

As previously shown, the design goal is to assign a diameter of column D and a spacing s_0 able to satisfy uplift equilibrium with adequate safety factors SF_{up}^* (Equation 7.16) and to allow no more than a tolerable water inflow Q_{tot} (Equation 7.18), if any.

If the whole plug volume is perfectly treated and the columns are overlapped to the maximum treatment depth, then $SF_{up}^* = SF_{up}$ (Equation 7.15) and $Q = 0$. Calculation of SF_{up}^* in the alternative case of incomplete overlapping may be performed, considering the filling ratio F defined in Figure 6.4, expressing the safety factor SF_{up}^* with the following relationship:

$$SF_{up}^* = \frac{\gamma_{jg}}{\gamma_w} \frac{\displaystyle\int_{h_{exc}}^{h_{exc}+h_2} F(z) \cdot dz}{\left(h_1 + \displaystyle\int_{h_{exc}}^{h_{exc}+h_2} F(z) \cdot dz \right)} \tag{7.19}$$

The variability of the filling ratio $F(z)$ with depth has been introduced to consider the divergence of contiguous columns.

For the equilateral triangular array, the worst condition can be conservatively assumed as the one producing the divergence between adjacent columns depicted in Figure 7.40 and by assigning cautious design values of the diameter D_d and deviation β_d. Under this assumption, the spacing $s(z)$ at any depth between two adjacent columns becomes

$$s(z) = s_0 + 2 \cdot z \cdot \cos\left(\frac{\pi}{3}\right) \cdot \tan(\beta_d) = s_0 + 1.73 \cdot z \cdot \tan(\beta_d) \qquad (7.20)$$

However, considering this situation as representative of the whole jet-grouted structure would certainly be overconservative, because the divergence of columns shown in Figure 7.40 would not be possible for all the cells. In fact, diverging columns (as in the figure) will have the opposite effect on some of the adjacent cells: at least for three of the twelve cells surrounding the three diverging columns, the effect on overlapping will tend to increase the filling ratio. For the rest of the surrounding cells, the effect of the divergence of the three columns is not so straightforward to predict, although it may be reasonably imagined that, on average, the nine cells will be affected in a minor way, and the effects on them will compensate.

As a consequence, an average filling ratio $F_{av}(z)$ can be introduced, considering a larger unit formed by the cell shown in Figure 7.40 and by the three surrounding cells positively affected by the divergence, assumed therefore to be completely treated. Figure 7.41 shows values of $F_{av}(z)$ for different conditions (columns spacing at ground level s_0, design diameter D_d and deviation angles β_d). Such values of $F_{av}(z)$ can then be used in Equation 7.19 to calculate SF_{up}^*.

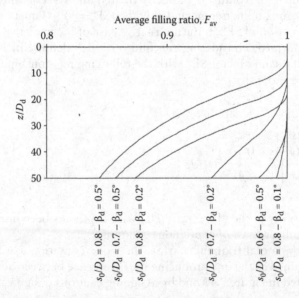

Figure 7.41 Deterministic calculation of the filling ratio for the equilateral triangular array.

For an excavation of prescribed depth and for a given water height above the depth of excavation (h_{exc} and h_1, respectively, in Figure 7.39), the deterministic design of the water sealing plug can be carried on with the following steps:

1. Fix design values of the columns' diameter D_d and inclination β_d.
2. Assume a tentative value of the column spacing at ground level s_0 able to guarantee water tightness of the plug for a prescribed thickness h_2^* ($F = 1$ for every $z < (h_{exc} + h_2^*)$ in Figure 7.41).
3. Consider the minimum safety factor $SF_{up,min}$ allowed by the Code of Practice of interest, and find the value of h_2 able to fulfil Equation 7.19. The integral in Equation 7.19 can be more easily computed as $F_m \cdot h_2$, taking F_m from Figure 7.41 as the mean value pertaining to the range of depths between h_{exc} and $h_{exc} + h_2$.

In case of imperfect water tightness of the plug [$F(z = h_{exc}) < 1$], the water flow in the unit cell can be computed by Equations 7.17 and 7.18, putting

$$A_{un,top} = (1 - F(h_{exc})) \cdot A_{cell} \tag{7.21}$$

7.5.3 Semiprobabilistic approach

With the semiprobabilistic approach, the design can be conducted without assigning arbitrary values of diameter and inclination but considering their probabilistic distributions. In this case, calculation can be performed by following the same procedure described in the previous section and by computing the average filling ratio with the charts reported in Figure 7.42 (for a 5% fractile), assigning the diameter D and inclination angle β by means of their mean values and variation coefficients.

7.5.4 Probabilistic approach

To make the design procedure more reliable, column defects in terms of diameter and inclination may be easily considered with the probabilistic approach, exploring different scenarios using the Monte Carlo method (see Chapter 6.4.5 and Figure 6.10), implemented to generate a large number of simulations to be analysed for extracting probabilistic information. This approach has been recently adopted, for instance, by Saurer et al. (2011) and Flora et al. (2012a).

In this section, results obtained with the Monte Carlo procedure for an equilateral triangular grid are presented in terms of simple design charts for some typical column diameters ($D_d = 1, 1.5$ and 2 m), different values of the relative spacing s/D ($0.66, 0.70, 0.80, 0.86$) and typical values of the

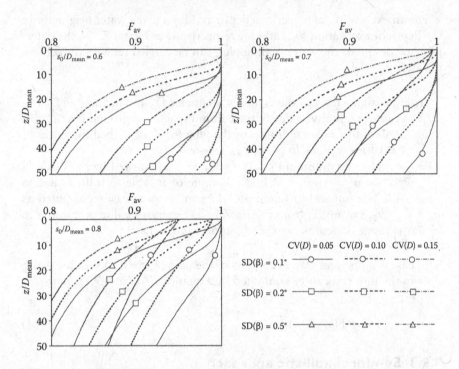

Figure 7.42 Semiprobabilistic calculation of the average filling ratio for the equilateral triangular array, to be adopted in the evaluation of the safety factor (Equation 7.19). The charts refer to the 5% fractile.

statistical parameters for a coarse-grained soil ($CV(D) = 0.2$, $SD(\beta) = 0.3$). Three excavation heights are considered ($h_{exc} = 5$, 10 and 15 m), with the water table at the most critical position (i.e., at ground level, $h_1 = h_{exc}$) and assuming a jet-grouted material unit weight $\gamma_{jg} = 20$ kN/m³. The results have been obtained by imposing a fractile of 5% and are plotted in Figure 7.43.

In the ideal case of perfectly vertical and cylindrical columns, the maximum ratio between spacing and column diameter to completely treat all the bottom plug volume is $s/D_{max} = 0.86$, which should be theoretically sufficient to obtain a bottom plug with the desired safety factors and with the minimum waste of jet grouting (overlapped volume). For larger spacing, even without defects, the barrier would not be impermeable because there would be a way for water to permeate through the untreated soil. Since columns always have defects, even for values of s/D lower than 0.86, the bottom plug may not be made of completely overlapped columns to the maximum depth of treatment. Once the diameter D of the column and the relative spacing s/D are given, the charts reported in Figure 7.43a, b and c can be used to assign the thickness of the bottom plug. $h_2 = z - h_1$

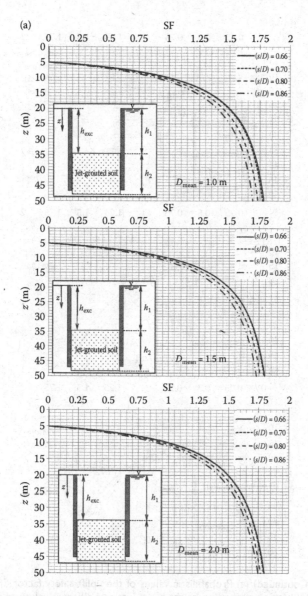

Figure 7.43 (a) Probabilistic values of the uplift safety factor SF (Equation 7.19) for an excavation height h_{exc} = 5 m and different values of the relative spacing s/D (CV(D) = 0.2, SD(β) = 0.3°, fractile = 5%). (b) Probabilistic values of the uplift safety factor SF (Equation 7.19) for an excavation height h_{exc} = 10 m and different values of the relative spacing s/D (CV(D) = 0.2, SD(β) = 0.3°, risk = 5%). (c) Probabilistic values of the uplift safety factor SF (Equation 7.19) for an excavation height h_{exc} = 15 m and different values of the relative spacing s/D (CV(D) = 0.2, SD(β) = 0.3°, risk = 5%).

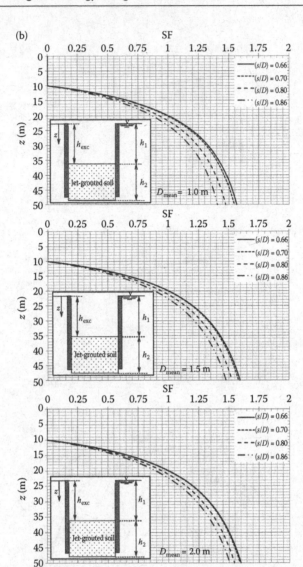

Figure 7.43 (Continued) (a) Probabilistic values of the uplift safety factor SF (Equation 7.19) for an excavation height $h_{exc} = 5$ m and different values of the relative spacing s/D (CV(D) = 0.2, SD(β) = 0.3°, fractile = 5%). (b) Probabilistic values of the uplift safety factor SF (Equation 7.19) for an excavation height $h_{exc} = 10$ m and different values of the relative spacing s/D (CV(D) = 0.2, SD(β) = 0.3°, risk = 5%). (c) Probabilistic values of the uplift safety factor SF (Equation 7.19) for an excavation height $h_{exc} = 15$ m and different values of the relative spacing s/D (CV(D) = 0.2, SD(β) = 0.3°, risk = 5%).

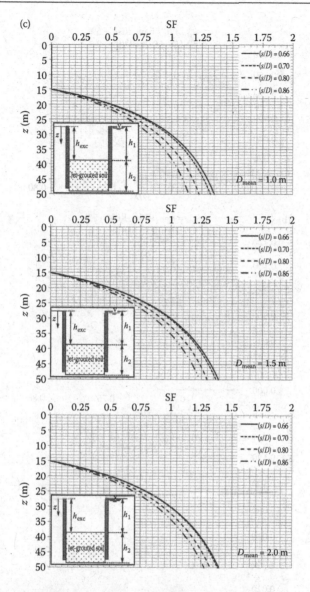

Figure 7.43 (Continued) (a) Probabilistic values of the uplift safety factor SF (Equation 7.19) for an excavation height h_{exc} = 5 m and different values of the relative spacing s/D (CV(D) = 0.2, SD(β) = 0.3°, fractile = 5%). (b) Probabilistic values of the uplift safety factor SF (Equation 7.19) for an excavation height h_{exc} = 10 m and different values of the relative spacing s/D (CV(D) = 0.2, SD(β) = 0.3°, risk = 5%). (c) Probabilistic values of the uplift safety factor SF (Equation 7.19) for an excavation height h_{exc} = 15 m and different values of the relative spacing s/D (CV(D) = 0.2, SD(β) = 0.3°, risk = 5%).

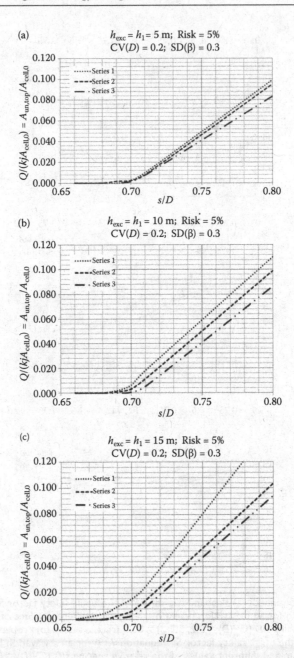

Figure 7.44 Probabilistic values (risk = 5%) of the water flow per unit cell (Equation 7.22) for different values of the relative spacing s/D and of the diameter D with the statistical parameters reported in the charts and for excavation heights. (a) h_{exc} = 5 m; (b) h_{exc} = 10 m; (c) h_{exc} = 15 m.

for the three considered reference excavation depths and for the desired uplift safety factor SF_{up}. It is pointed out that, for the considered statistical parameters, the maximum relative spacing to obtain a fully treated bottom plug ($Q = 0$) is $s/D = 0.66$, much smaller than the theoretical one ($s/D = 0.86$). However, the charts show that equilibrium can be granted even in the case of a bottom plug not fully treated, just making longer columns depending on the chosen value of s/D. For a given value of s/D, larger columns are more effective because, for all the values of $s/D > 0.66$, the thickness h_2 needed to have a given value of FS_{up} is smaller.

For these same cases, Figure 7.44 reports the water inflow Q into the unit cell of the bottom plug, for different values of the relative spacing s/D. For the sake of generality, in the diagram, the dimensionless ratio $Q/(A_{cell,0} \cdot j \cdot k)$ is reported, and the total inflow Q_{tot} (Equation 7.22) can be easily calculated from the diagram, multiplying the ordinate of the charts for the product ($j \cdot k \cdot A$), getting

$$Q_{tot} = \left(\frac{Q}{A_{cell,0} \cdot j \cdot k} \right) \cdot (j \cdot k \cdot A) \qquad (7.22)$$

For a given value of the relative spacing s/D, the charts show that larger diameters are more effective as they reduce the water inflow. Even for the most critical situation ($h_{exc} = 15$ m) the results indicate that, as long as uplift equilibrium is granted with a satisfactory safety factor SF_{up}, most times, the water inflow may be tolerated.

Chapter 8

Controls

8.1 OBJECTIVES OF CONTROLS

In general, the primary objective of control is to guarantee that products are able to meet the requirements for which they have been created. This is the basic principle of quality control (QC). Nowadays, the philosophy of quality is widely spread in all industrial fields and, in a very competitive market, it is considered a necessary tool to demonstrate the quality of the company and of its products. As a consequence, much experience has been gained in developing fast and effective testing procedures to be introduced in the industrial production chain. QC and quality assurance (QA) processes are usually codified by manuals prescribing a series of 'pass or fail' tests to be carried on at fixed steps of the production process. Their goal is to assess the quality of the basic components and the effectiveness of the working phases.

However, in civil engineering, the implementation of QC/QA procedures is more difficult mostly because the structures to be created are usually considered to be a sort of craftsmanship product rather than the result of an industrial process. In geotechnical engineering, QC may be particularly cumbersome because the products of the relevant industrial activities are normally buried underground. However, QC/QA is becoming an important issue also in geotechnical engineering, especially when ground improvement techniques are involved. Moreover, because there is still a relevant degree of uncertainty on the effects of soil treatment techniques, production and performance controls are often the best way to refine the decision-making process for subsequent projects.

The topic of control is widely dealt with in most geotechnical guidelines, specifying that a supervision of the construction process and workmanship by means of *in situ* and laboratory tests and the monitoring of the performance of structures during and after their construction have to be defined in the geotechnical design reports. Furthermore, there are applications, such as foundation piles or ground anchors, in which the role of experimental control becomes even more important, being codified at a very detailed level and considered as a compulsory step of the certification phase. When predictions are affected by a noticeable degree of uncertainty and the hazard associated with unexpected behaviours is particularly high, as may happen for complex geotechnical works, the so-called 'observational method' can be conveniently adopted, integrating control and monitoring activities in a closed loop design process.

The need for effective control and monitoring is particularly strong for jet grouting works, this technique being affected by three sources of uncertainty. The first one concerns the effectiveness of treatments, that is, the dimensions and properties of the jet-grouted elements; the second one pertains to the mechanical interaction between the jet-grouted elements and the surrounding soils, that is, to the overall behaviour of the jet-grouted structure; the third one is related to the possible undesired effects on the surrounding environment, with particular regard to the neighbouring structures, if any.

This chapter summarises the current practice regarding the experimental techniques used for the control of jet grouting execution and effects. The different control methods are presented according to the following main purposes:

- To ensure that the basic materials adopted for jet grouting possess adequate characteristics
- To check that the construction procedure is correctly carried out and that the equipment is working properly
- To quantify the dimensions and properties of the jet-grouted elements
- To verify the performance of jet-grouted structures
- To monitor the jet grouting effects on the surrounding environment and structures

It is worth emphasising how important and critical these tests may be because of the severe consequences that could arise from unforeseen defects of jet grouting treatments. For this reason, the site engineers should possess a deep knowledge of both the technological and the design aspects of jet grouting.

8.2 FIELD TRIALS

The so-called 'field trial' consists of carrying out preliminary jet grouting treatments and performing appropriate tests for the verification of the treatment results with respect to the design requirements. The goals of a field trial may be summarised as follows:

- To choose the best jet grouting system and select the most appropriate treatment parameters
- To assess the treatment results with respect to the project requirements

- To check the effects of treatments on the surrounding environment and structures
- To refine the control procedure to be implemented during construction

The field trial is thus a necessary step for linking the transition between the design and the execution of the jet grouting works. It is fundamentally aimed at verifying the effectiveness of the treatment before starting production, thus reducing uncertainties and preventing the occurrence of undesired effects. Examination of the field trial results may lead to the recalibration of the initially proposed solutions or, in some cases, even to the reformulation of the design process, in the spirit of the so-called 'observational method'.

Being a fundamental step for producing effective and safe jet-grouted structures, the field trials should be performed with the outmost care, highlighting all the relevant issues through appropriate testing. However, the significance of field trial results strongly depends on the similarity of geotechnical conditions, compared to the actual construction site. Therefore, the field tests should be carried out in the immediate vicinity of the work or at least in a similar geotechnical context.

In general, it would be desirable to conduct the field trial during the design phase. In fact, by applying such a procedure, the design quality would be certified, thus avoiding possible litigation that may subsequently arise between the client and the contractor. Nevertheless, for practical reasons, the field trial is usually carried out by the contractor in a preliminary construction stage. It is recommended, however, to plan the field trial at the design stage, stating the objectives, the extent and the method of the investigation to be performed.

Different jet grouting systems (single, double or triple fluid) and treatment parameter combinations may be tested during the field trials, with the aim of calibrating the execution procedure and pursuing cost effectiveness, given the specific design constraints. This goal can be achieved by producing prototype columns with different combinations of treatment parameters and by comparing the results. Whenever possible, the surrounding soil should be excavated after treatment to measure the column diameter and to check the continuity of the jet-grouted material (Figure 8.1).

Several kinds of tests can be then performed for evaluation of the geometrical and mechanical characteristics of single jet-grouted columns or column groups by applying a variety of techniques, as described in the following paragraphs. In any case, the properties of each prototype column (diameter, axis direction, compressive strength, etc.) should be measured by retrieving a sufficient amount of data to apply statistical criteria that are needed to quantify the variability produced by the coupling of each treatment procedure with the subsurface properties. Whenever possible, the overall performance of the jet-grouted structure (i.e., bearing capacity, water tightness) should be also tested.

Figure 8.1 Examples of field trials: (a) single-fluid jet-grouted columns obtained with different combinations of injection parameters in pyroclastic soils. (b) Double and triple jet-grouted columns obtained in sandy soils. (From Croce, P. and A. Flora, *Rivista Italiana di Geotecnica* 2: pp. 5–14, 1998; Mauro, M. and F. Santillan, Large-scale jet grouting and deep mixing test program at Tuttle Creek Dam. *Proceedings of the 33rd Annual and 11th International Conference on Deep Foundations*, Paper 1607: 10 p., 2008.)

The measurement of the possible undesired effects on the surrounding environment should be also investigated by field trials when jet grouting is performed in sensitive situations. In particular, a monitoring system should be conceived at the design stage and then tested during the field trial to detect possible ground movements that may be harmful to neighbouring structures. Pore pressure and temperature recording may be also useful to foresee possible environmental modifications.

Real-time monitoring, along with the registration of drilling and injection parameters, should be also pursued. The monitoring systems should be carefully tested and then used during construction to provide active QC by sounding alarm signals in case the prescribed threshold values of relevant variables are exceeded.

The examination of the field trial results is useful not only to calibrate the treatment procedure but also to verify that the adopted control devices and techniques are well suited to this aim. If possible, different monitoring systems should be compared at the field trial stage.

8.3 CONTROL RULES AND GUIDELINES

Although approached from different perspectives, the theme of QC is extensively dealt with in existing guidelines on jet grouting. As already mentioned in the previous chapters, the guidelines of the Japanese Jet Grouting Association (JJGA 2005) attach much importance to the design of jet-grouted structures and to the execution of treatments. They provide theoretical formulas and safety factors for the different jet grouting systems and applications, specifying the treatment parameters to be adopted for the different soil types. The role of controls is then devoted more to ensuring that the execution of treatments conforms to the treatment specifications than checking the final geometrical and mechanical properties of columns. The underlying idea is that the properties of columns are safely guaranteed if the construction specifications are fully respected (fluid pressure, flow rate, lifting speed, etc.). A detailed checklist is also provided (testing methods, test frequency, etc.) to certify the initial position of the machinery, the rotation and penetration rates of the monitor during drilling and lifting, the pressures and flow rates of the injected fluids (grout, air, and water), and the amount of returned spoil.

Following a different philosophy, the guidelines provided by the Grouting Committee of the Geo-Institute of the American Society of Civil Engineers (GI-ASCE 2009) leave greater freedom to the designer, who is not required to specify a set of treatment parameters. The execution of treatments, together with the plan to investigate their effectiveness, is totally left to the contractor under the strict supervision of the client's representative engineer. The contractor is fully responsible for the proper execution of jet grouting and must submit the results of the test to the client's representative engineer for approval of the proposed solutions. In this scenario, the controls have the fundamental role of optimising the execution of treatments, and thus, a fundamental step consists of performing a series of demonstrative field trials prior to the starting of works. Trial columns must be initially produced at a representative site, and their geometrical and mechanical properties must be measured to optimise the treatment procedure and to confirm that the jet-grouted material meets the design requirements. Once the treatment procedures are completely defined, controls during construction must be systematically carried out to make sure that the procedure has been correctly performed.

The ASCE guidelines also provide detailed specifications on materials (cement and additives), which must conform to the American Society for Testing and Materials (ASTM) or the American Association of State Highway and Transportation Officials (AASHTO) standards, as well as on the equipment characteristics (for drilling, grout mixing and injection), which must possess a proven suitability for jet grouting. Specific monitoring by means of automatic logging systems is prescribed to continuously measure the flows and pressures of all fluids, rotational speed, depth and the retraction rates of the rods. Continuous coring down to the full depth is then prescribed on a fixed percentage of jet-grouted columns, together with a detailed description of the procedures, to obtain samples of jet-grouted material, including sampling devices and curing boxes. All the monitoring and recording results must be presented in the form of tabular and graphical data to be submitted to the owner's representative engineer during the course of the work.

A rather similar approach is defined in the European Standard on the *Execution of special geotechnical works: Jet grouting* (EN 12716 2001). In particular, a list of activities is prescribed for jet grouting execution and control. With regard to QC, specific reference is made to the assessment of the geometrical configuration of the jet-grouted elements and to the properties of jet-grouted material (resistance, deformability and/or permeability).

According to the European rule, if jet grouting is performed in subsoil conditions similar to previous experiences (for which a detailed documentation is available), and if the static role of the jet-grouted elements is not critical for safety, the field trial can be omitted unless specifically requested in the project. In this case, however, the properties of the jet-grouted material should be measured on a relevant number of columns produced at the beginning of the work to check that the results meet the design requirements. In any case, continuous monitoring of the treatment parameters with periodically calibrated instruments and a record of the properties of returned spoil (density, Marsh viscosity, bleeding) should be carried out all over the production process.

Mechanical tests (unconfined compression, extension and shearing) on samples cored from the columns are finally prescribed with a specific frequency (e.g., four samples, each with 1000 m^3 of treated material). Checking the permeability is also required both for single columns, by performing pumping tests in holes previously drilled into the column, and for the whole jet-grouted elements made of several overlapped columns.

8.4 MATERIALS QUALIFICATION

The materials used for jet grouting must have a proven effectiveness for both the injection and the subsequent hardening of soil to guarantee adequate characteristics of the jet-grouted elements. However, a preliminary distinction should be made between the artificial components, such as cement, additives or reinforcing bars, which are produced by manufacturers and whose quality is certified by documents accompanying the products, and the water, which is usually taken near the construction site and must thus be controlled on site.

8.4.1 Cement

The cement–water suspension is the fundamental component of jet grouting. The characteristics of this slurry should allow injection without causing an excessive loss of energy during passage through pipes and nozzles,

Table 8.1 Main types of Portland cement[a]

Traditional classification		European classification (BS 8500-1:2006)
British	American	
Ordinary Portland (BS 12)	Type I (ASTM C150)	Type (CEM) I Portland
Rapid-hardening Portland (BS 12)	Type III (ASTM C150)	Type IIA Portland, with 6% to 20% fly ash, ggbs,[b] limestone or 6% to 10% silica fume
Low-heat Portland (BS 1370)	Type IV (ASTM C 150)	
Modified cement	Type II (ASTM C 150)	Type IIB-S Portland, with 21% to 35% ggbs[b]
Sulphate-resisting Portland (SRPC) (BS 4027)	Type V (ASTM C 150)	
Portland blast-furnace (slag cement) (BS 146)	Type IS Type S Type I (SM) (ASTM C 595)	Type IIB-V Portland, with 21% to 35% fly ash
High-slag blast-furnace (BS 4246)		Type IIB+SR Portland, with 25% to 35% fly ash with enhanced sulphate resistance
Portland-pozzolan (BS 6588; BS 3892)	Type IP Type P Type I (8PM) (ASTM C595)	Type IIIA+SR Portland, with 36% to 65% ggbs[b] with enhanced sulphate resistance
		Type IIIB Portland, with 66% to 80% ggbs[b] Type IIIB+SR Portland, with 66% to 80% ggbs[b] with enhanced sulphate resistance
		Type IIIC Portland with 81% to 95% ggbs[b]
		Type IVB-V Portland, with 36% to 55% fly ash

Source: Modified from Neville, A. M. and J. J. Brooks, *Concrete Technology*: Harlow, United Kingdom: Pearson Education, Ltd., 431 p., 1987.

[a] For American cements, the air-entraining option is specified by adding A; for ASTM C575 cements, moderate sulphate resistance or moderate heat of hydration or both, can be specified by adding MS or MH.

[b] ggbs, ground-granulated blast furnace slag.

and enable agglomeration with the remoulded soil and hardening to finally provide a water-resistant and chemically stable material.

Several types of Portland cement are available in relation to the required physical and chemical characteristics. Both the ASTM C150 (2002) Standard and the European classification group them into different categories, as shown in Table 8.1, each characterised by a number indicating its specific characteristics (hardening rate, heat development during hydration, aeration and resistance to sulphates, organic matter, salts).

Unless differently specified, ordinary Portland cement (Type I) of the strength class 32.5 is normally adopted; blast furnace or pozzolan Portland cement may be used when improved mechanical properties or low heat generated by hydration are specifically required.

Because jet grouting is an underground application, a nontrivial aspect for the selection of the cement type is represented by the resistance to sulphates. Concerning this issue, the U.S. Bureau of Reclamation (1981) provides the indications reported in Table 8.2, in which the best suited type of cement can be chosen depending on the sulphate concentration measured in the soil and water.

Cement is usually conveyed to the construction site in bulk form, and its quality is guaranteed by certificates issued by the supplier, which is responsible for executing controls within its production line. Additional controls are required when jet grouting is performed in specific environmental conditions to assess the compatibility of cement with the chemical characteristics of water and additives used to prepare the grout, groundwater, soil, etc. These controls can be performed by the manufacturer at the production site or by the contractor in the construction site. In all cases, laboratories should meet eligibility requirements in accordance with standardised rules (e.g., ASTM C1222 1999).

Table 8.2 Selection of the cement type on the basis of sulphate concentration in the soil and in the water

Relative degree of sulphate attack	Percentage of water-soluble sulphate (as SO_4) in soil samples	Sulphate (as SO_4) in water samples (ppm)	Cement type
Negligible	0.00–0.10	0–150	I
Positive	0.10–0.20	150–1500	II
Severe	0.20–2.00	1500–10,000	V[a]
Very severe	2.00 or more	10,000 or more	V+ Pozzolan[b]

Source: Modified from USBR, *Concrete Manual: A Manual for the Control of Concrete Constructions*, 8th ed.; Denver, CO: U.S. Bureau of Reclamation, 627 p., 1981.

[a] Approved Portland-pozzolan cement providing comparable sulphate resistance when used in concrete.
[b] Should-be-approved pozzolan that has been determined by tests to improve sulphate resistance when used in concrete with Type V cement.

8.4.2 Admixtures

The term 'admixture' refers to the complex of chemical products added to the cement–water grout at the mixing stage. For the qualification of the admixtures used to improve the properties of grout, the currently available standards for concrete (e.g., ASTM C33 2013) can be adopted.

If bentonite is used to improve stability or as a pumping aid, it is good practice to ensure that it has been prehydrated for some time prior to being incorporated in the mix. The indication of the ASCE Grouting Committee, in its Jet Grouting Guidelines (GI-ASCE 2009), is that this should be done for at least 12 h, unless demonstrated hydration is achieved in less time. Furthermore, the chemical–mineralogical composition of bentonite should be stated in the certificate accompanying the product. Typical controls on bentonite consist of the determination of the grain size distribution, consistency limits, pH, moisture content, settling time and Marsh viscosity.

Fly ash may be also added to improve the strength and durability of the jet-grouted soil. The class of this material according to the distinction made by standards should be firstly known to determine if it contributes to cementation only by reacting with $Ca(OH)_2$ to produce pozzolanic reactions (class F according to ASTM C618 2012) or if it has cementing properties on its own (class C according to ASTM C618 2012). The amount of silica and alumina in the fly ash, the presence of moisture and lime and the fineness and carbon content of the material are relevant for the activation of pozzolanic reaction. Therefore, these properties need to be known from the certificates released by the suppliers or determined with specific tests before the materials are used.

For other chemical additives (e.g., sodium silicates, barite, hematite, chlorides, magnesium and aluminium silicates), certifications are directly provided by the manufacturers.

8.4.3 Water

The water used for preparing the grout must possess characteristics that do not inhibit cement hydration and hardening. Drinkable water, or water coming from public nets, provided that it is free from smells and flavours, can be used for preparing the grout. Clear, non-corrosive groundwater or surface water can be alternatively used after having ensured that unfavourable agents (especially sulphates and chlorides) are not present in harmful percentages.

Because of the lack of rules explicitly referring to jet grouting, controls and limitations can be indirectly derived from the standards for concrete (ASTM C1602/C1602M 2012), in which the following limits are set for the water to be used:

- Temperature (<60°)
- Amount of sodium and magnesium chloride (<3%)
- Amount of sulphate (<6%)

- Amount of organic acids (>0.1%)
- Amount of organic matter and suspended clay (<2 g/L)

In particular cases, some of these constraints may be overlooked; for example, the presence of chlorides, usually limited in reinforced concrete to prevent corrosion of the steel reinforcement, may be tolerated in the case of nonreinforced jet grouting columns. All the limitations on water properties must be periodically verified at the construction site by laboratory tests.

8.4.4 Reinforcements

As shown in Chapter 6, the flexural, compressive and tensile resistance of the jet grouting columns may be increased by the insertion of steel or fibreglass frames in the freshly jet-grouted material or, after drilling, in the hardened column. The dimensions and mechanical properties of these materials are of primary importance for the effectiveness of reinforcements and thus should be accurately controlled before placement. Cross-sectional dimensions of the bars, yield and ultimate strength, deformation modulus and elongation of the material are properties that are normally provided by manufacturers and controlled after shipping. However, a visual inspection is recommended to ensure that reinforcements do not bear surface defects, such as small cracks or coating with materials, which may reduce their adhesion with the jet-grouted material. In the case of steel reinforcements, the surface should not be overly oxidised or corroded.

8.5 TREATMENT CONTROL

8.5.1 Grout preparation

The preparation of grout is accomplished in a plant that includes cement and water tanks, grout mixers, agitators, compressors and pumps, each with sufficient capacity to ensure adequate supply to the jet grouting monitors. It is customary to perform periodic calibration of the dosing equipment to ensure that the designed composition of the mix is respected. In any case, a series of tests should be periodically performed on the prepared grout to ensure that the mixture fulfils the necessary requirements, that is, to be injected without difficulty, to activate soil erosion and mixing and to develop the final required characteristics without suffering from the influence of external factors. To this aim, the cement content, density and viscosity of the fresh grout; the segregation of cement from water; the setting time; and the compression strength of the hardened mixture are factors to be controlled in the laboratory on samples periodically taken during production. The positive response from these tests represents an indirect confirmation of the proper functioning of the mixing plant and the quality of the basic materials (water, cement and additives).

Density and viscosity affect the ability of penetration into the soil of the mixture. A regular examination of the density, which can be performed with a Baroid balance (API 2009) or with a calibrated pycnometer, allows evaluation of the correct dosing of components.

As far as the evaluation of viscosity is concerned, it must be remembered that the grout is a suspension and not a solution. As such, it behaves as a Bingham fluid, that is, with a viscosity not depending on a single coefficient as for Newtonian fluids but dependent to a considerable extent on the instantaneous velocity. However, it is customary to define a conventional apparent viscosity evaluated by Marsh funnel testing (ASTM D6910 2009). Indicative values for cement–water mixes with variable cement contents are shown in Figure 8.2.

The loss of water can be quantified using the results of bleeding tests (ASTM C940 2010), which consist of pouring the mixture into a standard graduated recipient and measuring after a certain time interval the height of the segregated liquid fraction. Indicative values for different cement–water mixtures are provided in Figure 8.3.

It is also common practice to measure the compressive strength of the grout and its development with time. For a fast determination of the setting time, Vicat needle tests (ASTM C191a 2001) can be performed. Because the results of this test are strictly related to the water–cement ratio of the mix, delayed setting may indicate the malfunctioning of the mixing plant or the inadequate composition of the grout. To know the development of the compressive strength with time, samples may be formed with the

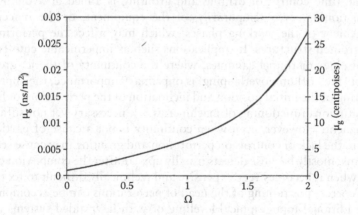

Figure 8.2 Dependency of the apparent viscosity coefficient μ_g on the composition of grout expressed in terms of the ratio by weight ω between cement and water (see Chapter 3). (From Raffle and Greenwood 1961, as reported by Bell, A. L., Jet grouting, In M. P. Moseley, ed., *Ground Improvement*: Blackie: pp. 149–174, 1993.)

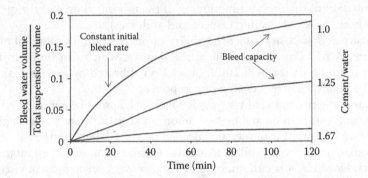

Figure 8.3 Bleeding of ordinary cement–water mixtures, with different water–cement ratios. (Modified from Littlejohn 1982, as reported Bell, A. L., Jet grouting, In M. P. Moseley, ed., *Ground Improvement*: Boca Raton, Florida: Blackie, pp. 149–174, 1993.)

material coming from the mixing plant, carrying out uniaxial compression tests (ASTM C109 2012) at different times after the completion of hardening. However, these results are only indicative of the development in time of the jet-grouted material strength and cannot be used to quantify its final value, which depends on the interaction with soil.

8.5.2 Drilling and grouting

The real-time control of drilling and grouting is aimed at avoiding the malfunctioning, even temporarily, of the equipment, or the inaccurate management of the working phases, which may affect the performance of jet-grouted structures. In applications such as impermeable cutoffs and bottom plugs or tunnel canopies, where the continuity of the jet-grouted elements (i.e., column overlapping) is of primary importance, the appropriate setting of the initial position and inclination of the perforation tools and the continuous functioning of machineries is a necessary yet not sufficient requirement. However, even when continuity is not a cause of particular concern, the lack of controls on perforation and grouting may cause serious problems, mostly because defects usually appear after the completion of the work, when it becomes very expensive and technically difficult to fix them.

The correct positioning of the head of perforations can be accomplished with traditional topographical levelling or with laser-aided systems referring to an absolute or relative system of benchmarks. Figure 8.4a shows the unusual case of jet grouting works accomplished to seal the foundation of a river weir: in this case, an impermeable cutoff has been made by the overlapping of jet-grouted columns produced by working from a floating barge. The initial position of the columns was defined above the water level on a

(a)   (b)

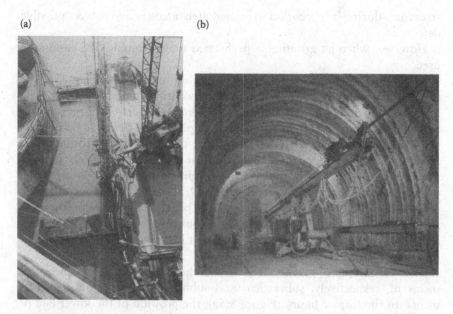

Figure 8.4 Positioning of the head of perforation on a floating barge (a) and on the excavation front of a tunnel (b).

chain fixed to a system of arms anchored to the weir. For the jet-grouted
tunnel canopy shown in Figure 8.4b, the correct positioning of the drilling
mast was determined by placing a reference template on the back of the
equipment and by sighting it from the apex of the cone frustum.

These positioning methods can be conveniently adopted when the position of all the columns is defined in a unique phase before performing the
treatments, but they become rather cumbersome when topographical measurements have to be routinely alternated with treatments. For this reason,
Global Positioning Systems (GPS) have progressively become more popular
in recent times, thanks to a faster and easier use, and also to the possibility
of automatically recording measurements. Open sea treatments are probably the most enlightening examples in which GPS can be conveniently
applied. In the absence of topographical references, drilling tools may, in
fact, be guided with a precision of a few millimetres by a couple of GPS
aerials and a clinometer mounted directly on the drilling mast. However,
because the measuring device is placed at a height of some metres on the
machinery, GPS accuracy strongly depends on the correct verticality of the
mast. In sites with uneven ground, it is not uncommon to see largely erroneous positioning caused by nonvertical masts.

The processing with dedicated software of these and other data, such
as inclination and the depth of the drilling bit and injection parameters,
allows building in real time a three-dimensional model of the jet-grouted

structures during its execution to immediately identify and correct possible defects.

However, when jet grouting is performed underground, GPS cannot be used.

The inclination of the drilling rods is another fundamental characteristic to be measured in those applications where continuity must be ensured by an effective overlapping of columns or where the direction of perforation must be carefully governed to follow complex paths. Nowadays, different systems have been developed to measure the direction of ground holes; some of them, equipped with pendulum or gyroscope, provide a direct measurement, whereas an indirect measurement of the inclination of the drilling rods can be given by monitoring with aerial GPS the position of the top end of the drilling mast. In all cases, the problem with the current measurements of axis deviation is that the true position of the column is known only after complete rod retrieval, and thus, it is not possible to correct it on real time during execution.

The two examples given in Figure 8.5 show the perforation measurements of, respectively, subvertical and subhorizontal jet grouting treatments. In the former figure (Figure 8.5a), the position of the lower end of 61 columns forming a vertical cutoff is reported. Measurements have been performed at a depth of approximately 37 m from the head of perforations, and dots are reported with reference to a system centred on the ideal position of the column axes. A rapid checking of data shows a maximum deviation of 33 cm (correspondent to 0.9% of the column's length). The distinction between systematic and random effects clearly shows an average heading of all columns toward the northwestern direction. However, it must be considered that this effect could not be the most severe for the continuity of the cutoff because it does not affect in a critical way the parallelism of

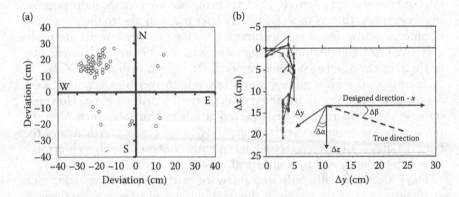

Figure 8.5 Examples of misalignments of subvertical (a) and subhorizontal (b) columns.

columns; in this sense, more troublesome consequences may occur because of random deviations.

Figure 8.5b shows the position of the end of five subhorizontal columns, each approximately 12 m long, forming a part of a tunnel canopy. Deviation, in this case, reaches a maximum of 20 cm (1.7% of the column's length). This systematic downward deviation of columns, caused by the self-weight of the drilling equipment and systematically reported in canopy execution, while not being particularly troublesome for the continuity of the canopy, may cause a deviation of the subsequent sectors of the tunnel and, therefore, must be corrected.

All the equipment used for drilling boreholes, and for lowering, retrieving and rotating monitors should be of a type and capacity suitable to perform the work as shown on the design drawings. In modern machinery, these actions are continuously monitored throughout execution by an automatic acquisition system that includes a set of periodically calibrated measuring instruments (pressure gauges, displacement transducers, load-measuring devices), a data logger and software to process data and to plot them on a computer screen. A variety of solutions is available on the market, which can be customised to the specific application by selecting the most appropriate set of variables. A typical list of drilling and injection variables is reported in Table 8.3, and a representative log of treatment parameters is reported in Figure 8.6 for triple-fluid jet grouting.

The reports that can be obtained with these systems, in which each monitored parameter is plotted versus the distance from the head of excavation, are useful to confirm the regular execution of the jet grouting columns to detect the occurrence of singularities, such as the presence of cavities or hard soil strata, and to warn about the momentary malfunctioning of the injection equipment (fall of pump pressure, nozzle obstruction, etc.), which may affect the treatment result.

During construction, it is also important to control the amount, continuity and quality of the spoil returning to the borehole's head. In fact, it is of outmost importance to ensure that the soil is remoulded by the jet

Table 8.3 List of parameters automatically recorded during the execution of jet grouting

Drilling	Grouting
Advancing rate	Withdrawal speed
Rotary speed	Rotary speed
Tool pressure	Air pressure
Torque	Water flow rate and pressure
Drilling mud flow rate and pressure	Grout flow rate and pressure
	Injected volume of grout

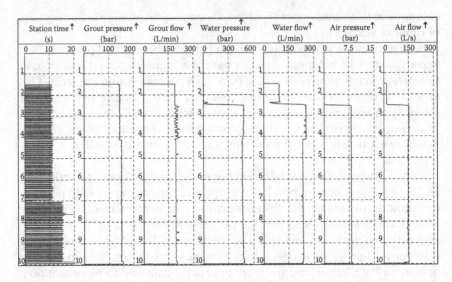

Figure 8.6 Representative log of treatment parameters for triple-fluid jet grouting. (Courtesy of Trevi.)

kinetic action and not fractured by static fluid pressure, as already discussed in Chapters 3 and 4. In fact, if the latter phenomenon occurs in the hole, even temporarily, the columns may have a shape quite different from the ideal cylindrical one. To avoid this problem, the spoil should continuously flow through the borehole and not follow different paths. Absence of spoil return or spillage from other paths at the ground surface should alert the operator to the occurrence of jet malfunctioning. Any anomaly of the flow outcoming from the borehole should, therefore, be promptly noticed.

In general, the amount of spoil returning at ground surface decreases as grain size increases and may become nil in clean gravels, in which the whole injected grout flow can sometimes be absorbed into the very large pores of the soil. Therefore, the existence of adsorbing clean gravel strata should be detected before carrying out treatments to avoid misinterpretation of a poor spoil outflow.

The composition of spoil return may be also indicative of the grout–soil interaction phenomenon occurring into the soil. In fact, because the specific weight of the grout at the usual water–cement ratios (~1) is significantly lower than that of soils, a measurement of the spoil density can give an idea of the mechanisms occurring at depth. For instance, a spoil with a density similar to that of the injected grout may be indicative of a reduced erosive capacity of the jet; on the contrary, if erosion is occurring, the density of the spoil is in between that of the slurry and of the soil. In sandy materials, the density of the spoil has been observed to decrease with depth (Croce and Flora 1998), in accordance with the relationship between the resistance of soil to erosion and the stress level. Spoil density variations during jet grouting can be also adopted as operational controls in unconventional treatment procedures. This is the case of jet grouting executed from a certain depth during drilling (Sanella 2007; Flora and Lirer 2011) to save on cement use (see Chapter 2).

In massive treatments, when columns are overlapped, the density of the spoil decreases from primary to secondary or tertiary columns. Because a decrease of spoil density means a less effective eroding action, when a sharp decrease of density is observed, it could be convenient to change the mutual position of the columns to reduce shadowing effects and to make jet grouting more cost effective.

It is finally worth considering that, because of all these differences in composition between the spoil and the *in situ* jet-grouted material (Croce and Flora 1998), mechanical tests carried out on samples formed with the spoil are not completely reliable and may give only a rough indication of the real properties of jet-grouted soil.

8.5.3 Spoil return

During the grouting stage, spoil return is probably the most important QC indicator on site. Therefore, it should be visually monitored during production, and flow characteristics should be carefully controlled. Unexpected reductions in spoil return may be indicators of a critical and undesired malfunctioning of the jet action (see Chapter 3). In case of negligible spoil return, it is compulsory to check that there is no clogging of the borehole annulus. If clogging is feared or observed, the production procedure has to be revised either by changing the jetting parameters or by modifying the drilling procedure using casing or bentonite slurry to sustain the borehole (see Chapter 2).

In any case, for the ease of control and to avoid environmental problems, spoil has to be channelled to storage, which is either a tank or a pond. Because the disposal of the spoil is a concern, it is usually removed as industrial waste, sometimes with very restrictive prescriptions. As a consequence, there is a recent tendency to reuse spoil, if possible, to strengthen landfills or even to inject new jet-grouted columns. The latter option has been recently proposed by Yoshida (2012), who indicate, however, that a more complex pumping layout has to be adopted in this case. As a consequence, the reuse of the spoil to create new columns may be cost effective only if spoil disposal is extremely expensive. An interesting outcome of spoil reuse is the global reduction in CO_2 emission, considering the overall life cycle of the cement used for jet grouting production (Yoshida 2012).

What is important to note here is that, if spoil has to be reused, its composition must be carefully checked, and controls must be more severe than usual.

8.6 PROPERTIES OF THE JET-GROUTED STRUCTURE

The inspections carried out after the execution of treatments are aimed at ensuring that the characteristics of the jet-grouted elements comply with the design requirements and that elements are able to contribute to the expected overall response of the jet-grouted structure. According to common practice, these controls can be subdivided into the following distinct but complementary categories:

- Control of the geometrical and mechanical characteristics of the jet grouting elements
- Control of the performance of jet-grouted elements and structures

For some applications, such as waterproofing barriers, these two aspects are so tightly interconnected that controlling the former is implicitly equivalent to verifying the second.

A positive result of these controls provides answers to a number of questions. First of all, it is a confirmation of the expected geometrical and mechanical properties of the jet-grouted elements, which is important feedback to confirm that the adopted injection system and the set of treatment parameters are well suited to the case. Then, it confirms that the model (empirical, semiempirical or theoretical; see Chapter 5) originally adopted to predict jet grouting effects, is reliable.

Controls of both the aforementioned kinds (characteristics and performance of jet-grouted structures) may be performed with different goals, depending on the phase of the production process, which can be schematically divided into the design stage, the first field trials, construction monitoring and refining and the final result check. For very important works, and in the case of long-term construction processes, it is customary to prolong the field trials during the first steps of execution by using controls to test production and to eventually readjust something in the treatment procedure.

In planning these tests and in processing results, it is finally worth remembering that results may be affected by significant variations, and thus, it is convenient to refer to statistical models for their interpretation (see Chapters 4 and 6).

The determination of the geometrical and mechanical characteristics of jet-grouted elements can be pursued with different techniques. Each of them is able to reach a certain degree of accuracy but, on the other hand, is often affected by a non-negligible level of invasiveness. The description of the available techniques, and of their effectiveness and their limits, is the main goal of this section.

8.6.1 Diameter of columns

The measurement of the diameter of columns is, by far, the most important control to be carried out on jet-grouted columns. The bad practice of considering the columns as piles of a given diameter is completely wrong, as largely discussed and shown in all the previous chapters, and as a consequence, it is compulsory to measure on site both the average diameter and its variations.

A variety of inspection techniques has been developed to measure the diameter of columns. Because each method is characterised by a degree of accuracy but may also influence the performance of the investigated element (for instance, boreholes may affect water tightness), comparative evaluations should be performed to decide which technique is the most suitable for the analysed case.

The measurements can be direct (in the sense that the diameter is directly measured) or indirect (in the sense that variables other than the diameter are measured to get information on the diameter). The most complete and accurate results are certainly obtained by directly measuring the cross-sectional dimensions throughout the columns, but this is not always feasible.

The ideal direct method, at least in field trials, is the discovery of columns (Figure 8.7). There is no better way to have complete information on the columns because, by directly looking at them, it is possible to highlight even the smallest but possibly crucial detail in terms of defects. Statistically representative samples of diameter measurements can be also obtained in this case, and relationships can be established between mean dimensions and injection parameters, which are of great help both in the specific site and in increasing the database for the calibration of predictive methods (e.g., Figure 4.5).

Obviously, since this method is particularly invasive, it can be planned and carried out only on trial columns that have no relevant role for the future structure. Furthermore, for cost effectiveness, it can be accomplished only to a limited depth and above ground water level, unless well points are temporarily adopted to lower the water table. Because the most superficial part of the columns is necessarily discovered, it may represent the best possible result in terms of dimensions. In fact, because of the general increase of soil shear strength with depth, a reduction of the diameter with depth is possible (see, e.g., Figure 4.2). Therefore, a direct observation of vertical jet-grouted columns is a good way to measure the diameter only if the jet-grouted structure is superficial too; otherwise, the field trial

Figure 8.7 Field trial: (a) examples of column exposure and (b) direct measurement of the diameter of discovered vertical columns.

may lead to a dangerous overestimation of the diameter (assuming that the injection parameters are kept constant with depth; see Chapters 2, 3 and 4). If columns have to be created at depths much larger than the ones possibly inspected, other inspection techniques must be preferred. For these reasons, a direct observation of columns can be performed only in a few cases, mostly with preliminary trial purposes.

A different application, in which this method can be routinely applied during construction, is the one reported in Figure 8.8, describing the measurement of diameter in trial horizontal columns executed within the cross-section of a tunnel under construction. In this case, the field trial can be performed with minimum cost, and measurements on the trial columns may be used to continuously control the effectiveness of performed treatments, also considering possible variations in soil composition and allowing the adjustment of the injection procedure for the further steps of excavation.

A less invasive direct technique, which is not widespread but is promising, is based on the use of a calliper (Figure 8.9a) having two arms that can be opened by increasing the pressure within a central hydraulic jack (Langhorst et al. 2007). The measurement consists of inserting the tool at different levels into the freshly injected column, varying the pressure of the

Figure 8.8 Measurement of the diameter of horizontal trial columns at the excavation front during tunnelling.

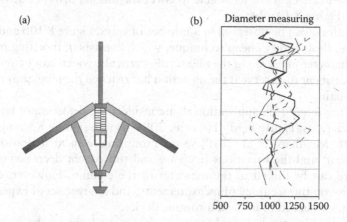

Figure 8.9 Calliper for the measurement of column diameter (a) and typical results (b). (Modified from Langhorst, O. S. et al., Design and validation of jet grouting for the Amsterdam Central Station. *Geotechniek*, Special Number on Madrid 14th ECSMGE: pp. 20–23, 2007.)

fluid contained in the jack and measuring the corresponding variation of volume in the jack chamber. The pressure–volume relationship is sensitive to the resistance offered by the surrounding material to the expansion of the two arms, and thus, a sharp deviation from the previous trend, with an increase in resistance, is noticed when the arms reach the undisturbed soil, that is, the sides of the column. The measurements reported in Figure 8.9b shows repetitive results as a proof of the efficiency of the system. Furthermore, it can be used to measure the diameters of the columns of the structure, being, therefore, in principle, a tool to carry out controls during construction. However, there is certainly a problem related to the difficulty of inserting the calliper exactly along the jet-grouted axis: a misalignment would result in a measurement that is likely smaller (of an unknown quantity) than the true diameter.

All direct measurements require time and may interfere with jet grouting operations. Therefore, they are carried out on a limited number of columns, most of the times prior to job execution, with the main goal of calibrating the jet grouting procedure, and cannot be considered routine control techniques.

The indirect methods, which allow information to be obtained on the diameter by the observation of other parameters, are the most common ones and can be used both in field trials and during job execution.

Examples of indirect measuring techniques based on the observation of the effects of jet grouting within inspection holes placed at different distances and parallel to the centreline of columns are reported in Figure 8.10. The monitoring holes are positioned at variable distances around the injection hole to check if the jet is able to cover the mutual distance at variable depths (Figure 8.10a, b).

The effects can be observed in a number of ways. Figure 8.10b and c, for instance, show a very cheap technique, which consists of inserting painted small diameter pipes along the measuring verticals, which can be retrieved after treatment to observe if the jet action has reached the pipe, thus removing the paint.

Another interesting application that considers the detection of temperature variations (Figure 8.9d) (Ho et al. 2001, as reported by Katzenbach et al. 2001; Meinhard et al. 2010) seems promising: cement hydration produces heat, and the variations (increase and subsequent decrease) of temperature can be linked to the diameter of the column. However, doubts may arise on the accuracy of measurements, and a larger set of experimental data are needed to make it a routine device.

More sophisticated controls can be also carried out in the holes. For example, an interesting method consists in registering the noise generated by the impact of the jet on one or more pipes (Figure 8.10e). With this method, the pipes are filled with water, and the noise is recorded by some hydrophones (Langhorst et al. 2007). The amount of energy emitted from the passing jet is transformed into an analogue electrical signal, whose power allows an estimation of the distance between the hydrophone and the jet. If carefully calibrated on site, this method may, in principle, allow a measurement of the diameter even if the jet does not directly reach the pipe, at least to a certain distance to be evaluated case by case. In all these methods, it is fundamental to have inspection holes parallel to the columns or, better, to know their position with sufficient precision; otherwise, measurements will be affected by significant (and unknown) errors, which would result in only qualitative information. To this aim, a measurement of their inclination is recommended, which, however, makes the measurement

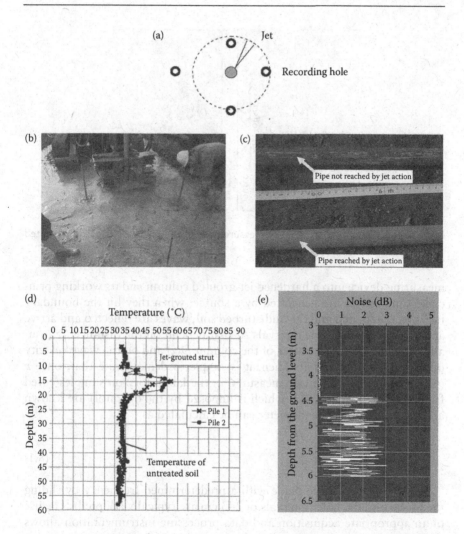

Figure 8.10 Measurement of column diameter from inspection holes (a) geometric scheme; (b) and (c) visual inspection; (d) temperature; (e) noise recorded by hydrophones.

more expensive. Because of this, these measurement techniques cannot be used for a routine check of the treatment execution but can only be implemented in field trial tests or for an occasional control to be performed during construction.

A rather popular indirect technique is the so-called 'sonic logging test' (ASTM D5753-e1 2012) carried out into a borehole created along the axis of the hardened column. Figure 8.11 reports the scheme of a sonic

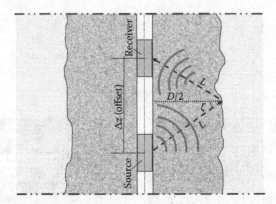

Figure 8.11 Scheme of sonic indirect measurement of the diameter D of a jet-grouted column.

measuring device into a hardened jet-grouted column and its working principle: sonic waves are generated by a source; when they hit the boundary between the treated and the undisturbed soil, waves are reflected and arrive at the receiver after time intervals Δt that depend on the distance Δz (usually of ~1 m), on the diameter of the column $D(z)$ and on the wave velocity (v) in the jet-grouted soil. Then, it is simple to demonstrate that, once v (which has to be previously measured in the laboratory on samples cored from the borehole) and Δz (which is imposed by the operator) are known and Δt is measured, the diameter can be calculated as

$$D = \sqrt{(v \cdot \Delta t)^2 - \Delta z^2} \tag{8.1}$$

Measurements can be made with a predetermined frequency by sliding the probe into the hole (intervals of 2–3 cm are typically adopted). The use of an appropriate acquisition and data processing instrumentation allows the return of almost continuous information along the column axis.

Using this working principle, more refined measurements can be obtained, carrying out a tomography of the physical properties of the jet-grouted column. Whatever the generated signal (electric, acoustic, etc.), the common principle is that a wave is triggered from a source and recorded by a sequence of receivers, all placed within the same borehole placed at the centre of the column. Once a large number of measurements have been carried out in the single column from a number of different positions, inversion algorithms are used to calculate a field of the measured property and, therefore, the dimensions of the column.

The example of electrical resistivity tomography reported in Figure 8.12 shows the capability of the method.

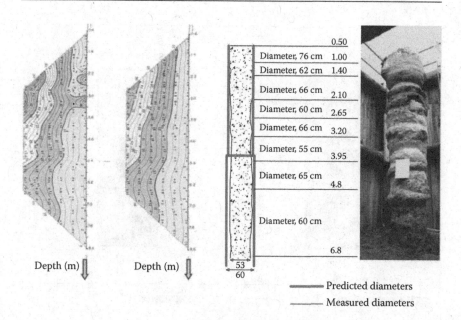

Depth (m)   Depth (m)

Diameter, 76 cm	0.50
Diameter, 62 cm	1.00
	1.40
Diameter, 66 cm	2.10
Diameter, 60 cm	2.65
Diameter, 66 cm	3.20
Diameter, 55 cm	3.95
Diameter, 65 cm	4.8
Diameter, 60 cm	6.8

——— Predicted diameters

——— Measured diameters

Figure 8.12 Measurement of column diameter with electrical resistivity tomography. (From Arroyo, M. et al., *Informes Sobre Tratamientos de Jet Grouting.* ADIF LAV Madrid-Barcelona-Francia, Tramo Torrasa-Sants. Report of the Universidad Politecnicha de Catalunya [in Spanish], 110 pp., 2007.)

A much more reliable result can be obtained if the tomography is carried out using more than one borehole placed outside the treated volume. Figure 8.13 shows the example of a single jet grouting column executed in a field trial, surrounded by four boreholes equipped with chains of sensors. These instruments receive dynamic signals (shear waves having velocity v_s) from a source alternatively lowered at different depths in all boreholes. The figure shows the very nice result of such a kind of tomography, which enables spotting out even minor details on the effect of jet grouting. The different values of v_s within the column indicate a nonhomogeneous composition of the jet-grouted mass, thus also giving a very useful extra piece of information. Usually, consistently with the physical interaction mechanisms expected between the soil and the grout and with the indications reported in Chapter 4 (see Chapter 4.4 and Figure 4.14), the top part of the column has a larger cement content and, therefore, larger values of v_s and better mechanical properties. Actually, the information on jet-grouted mass composition is not a minor outcome of multiborehole tomography, as will be shown in the following on groups of columns.

Topographic surveys be adopted for routine investigations and are usually adopted as spot-like controls to be carried out during field trials or, occasionally, during construction to confirm expected results.

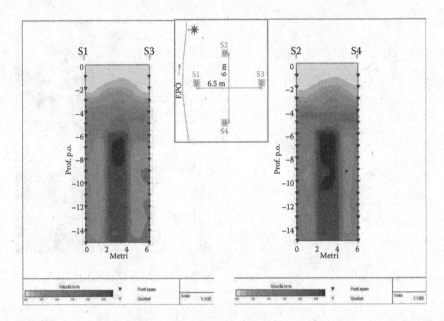

Figure 8.13 Example of seismic tomography on a single column. Different shadowing colours within the column indicate a nonhomogeneous composition of the jet-grouted soil (darker parts indicate higher values of v_s and, therefore, higher cement content). (From Ciufegni, S. et al., *Strade and Autostrade* 4-2007: pp. 156–162 [in Italian], 2007.)

Some authors (e.g., Morey and Campo 1999) claim that spoil density measurement is a useful indicator of the amount of treated volume and, therefore, an indirect means to estimate the mean diameter of the column. The topic of spoil composition has been dealt with in Chapter 4.4, where it has been shown that its composition may vary during the jetting action, the amount of removed soil and retained grout being variable with depth. In Chapter 4, however, quantitative evaluations of the removed soil or retained grout could be carried out because experimental information on the diameter had been previously obtained on columns exposed in a field trial. The other way around (i.e., calculating the diameter from spoil density measurement) is not possible with a rigorous calculation because only a mass balance is possible (under very restrictive hypotheses), which is not sufficient to have information on the treated volume (and, therefore, on the column diameter). Therefore, this procedure must be considered as fully empirical and can be adopted with extreme care and only if tuning has been carried out in a field trial.

Another indirect method, probably the most used one in routine controls, is the execution of boreholes to check the composition of the retrieved samples. By drilling boreholes at different distances from the column axis, information on the average diameter and on possible defects of treatment can be observed. The degree of accuracy of these indirect controls is very poor: most details are unavoidably lost and often, when the hole is bored just on the border of the column, it tends to deviate toward the less resistant untreated soil, thus giving erroneous geometrical information. However, it has the great advantage of being a very simple and cost-effective.

8.6.2 Continuity and homogeneity of jet-grouted elements

When columns are closely spaced to form continuous two- or three-dimensional bodies of jet-grouted material, the continuity and quality of cementation must be carefully controlled because they are of primary importance for the success of treatment. A preliminary control of the relative position and of the diameter of contiguous columns is, therefore, important.

Different tests are available to this aim. In field trials, continuity is usually controlled, creating some columns at the design relative spacing and checking the effectiveness with simple boreholes drilled at the intersections between columns (Figure 8.14).

The example of Figure 8.14 shows that the observation of the cored material, distinguishing the treated from the untreated parts, gives a clear picture of jet grouting effectiveness. In the case shown in the figure, the borehole farther from the column's axes (borehole A3) is the one showing the worst treatment effectiveness, with a complete lack of treatment from a depth of a few metres.

In some cases, especially if the boreholes are drilled in a freshly injected column, it may be difficult to univocally verify the effectiveness of the treatment. For example, cohesionless samples retrieved from the borehole may be erroneously interpreted as being untreated, whereas it may be the case that hardening has not occurred yet in the treated volume.

To move from a purely qualitative to a quantitative evaluation, the degree of cementation of the cored material should be somehow defined. A first, immediate way to do it is through the well-known Rock Quality Designation (RQD) Index (ASTM D6032 2008), that is, calculating the percentage of cored pieces of a sample longer than 100 mm. In the example shown in Figure 8.15, in which this value has been calculated for partial lengths of 1 m on three boreholes drilled in a massive foundation block, the presence of weaker portions of the cemented material can be noticed.

Another index used to quantify the effectiveness of cementation is the Core Recovery Index (Yoshitake et al. 2003) described in Figure 8.16. Although based on a qualitative description, a mark (from 1, very good, to 6, very poor) is introduced to quantify the level of ground improvement.

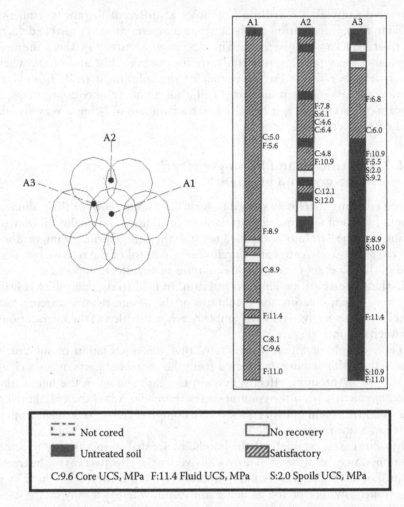

Figure 8.14 Assessment of the continuity of a jet grouting rosette by coring. (From Croce, P. et al., *Jet Grouting: Tecnica, Progetto e Controllo*. Benevento, Italy: 221 p [in Italian], 2004.)

Instead of coring samples from boreholes, which may be costly and time consuming, instrumented drilling represents a faster and no less effective alternative. This technique (ASTM D5434 2012), which is typically used for rocky materials (Brown and Barr 1978), consists of continuously record-ing the main variables (the rotation and advancing speed of the tool, the applied thrust and torque, the pressure of the perforation fluid) through-out perforation, which is fast because no samples are retrieved. Because this measurement can be applied to almost any equipment, the method is generally rather quick and inexpensive when compared, for example, with

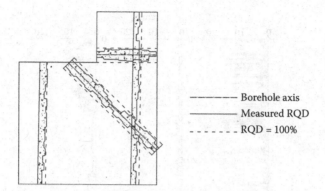

Legend:
- - - - - - Borehole axis
———— Measured RQD
- - - - - RQD = 100%

Figure 8.15 RQD values calculated from boreholes drilled in a massive foundation block.

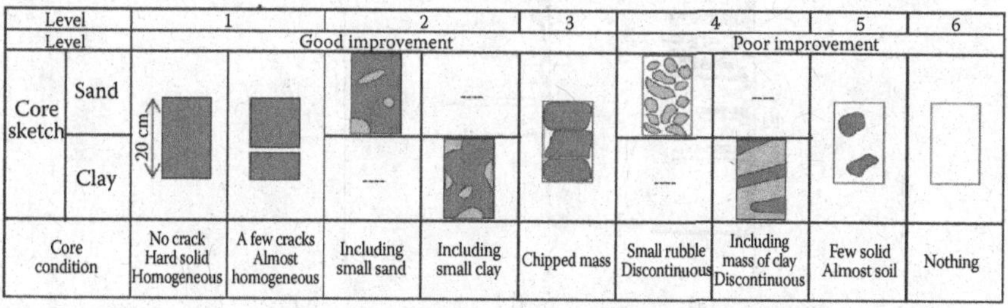

Level		1		2		3		4		5	6
Level		Good improvement						Poor improvement			
Core sketch	Sand										
	Clay										
Core condition		No crack Hard solid Homogeneous	A few cracks Almost homogeneous	Including small sand	Including small clay	Chipped mass		Small rubble Discontinuous	Including mass of clay Discontinuous	Few solid Almost soil	Nothing

Figure 8.16 Core Recovery Index of samples cored from jet grouting. (From Yoshitake, I. et al., *Proceedings of the Japan Society of Civil Engineers* 735: pp. 215–220, 2003.)

continuous sampling. It should, however, be noted that the derived information is mainly qualitative, especially because of the lack of universally accepted interpretation criteria. The example reported in Figure 8.17 shows noticeable differences between the advancing velocities of a drilling bit penetrating in a virgin soil, in a successfully jet-grouted soil and in a material with a variable degree of cementation.

Various dynamic tests can be also used to determine the homogeneity of cementation in jet-grouted bodies. Their basic principle consists of determining the propagation velocity of primary and secondary waves travelling through a definite space. The propagation velocity of waves is determined, knowing the distance between the source and the receiver, by measuring with an oscilloscope the time to cover this distance. Assuming the material equivalent to an elastic medium, the speed of wave propagation can be correlated to the stiffness of the material in the examined portion. Particular care is recommended when this technique is applied using compressive wave velocity in fully saturated materials because water may affect the results.

The previously described sonic logging tests (ASTM D5753-e1 1995) can be also used. A typical representation is reported in Figure 8.18, in which

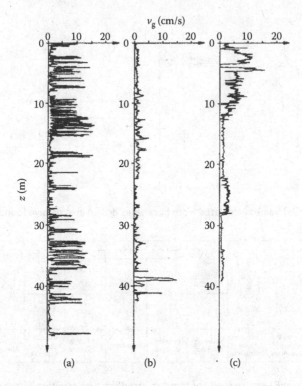

Figure 8.17 Advancing speed of the drilling tools in a virgin soil (a), in a jet-grouted material (b) and in a soil with a variable degree of cementation (c). The sharp reduction in velocity is a clear mark of cementation.

the white and the black shadowed areas represent, respectively, the positive and the negative half waves recorded at various depths, which are plotted in the time domain. The shape of the waves highlights the regularity of the profile in the lower part (the light and dark stripes are regular and parallel, indicating a homogeneous material), with good mechanical characteristics of the treated soil (the compressive wave velocity v_p is ~5000 m/s). Conversely, inflections, confused trends or missing strips in the upper part of the column indicate a worse and less homogeneous material (the compressive wave velocity v_p is equal to 4000 m/s).

In another example of sonic logging tests (Croce et al. 1994), measurements have been carried out along six vertical boreholes drilled in a cylindrical jet-grouted shaft (Figure 8.19a, b). The heterogeneity of the material is clear from the results summarised in Figure 8.19c. Four different areas could be, in fact, distinguished within the jet-grouted element: in the upper part, for a thickness of approximately 1 to 2 m, the compressive wave velocity v_p decreases toward the ground surface because

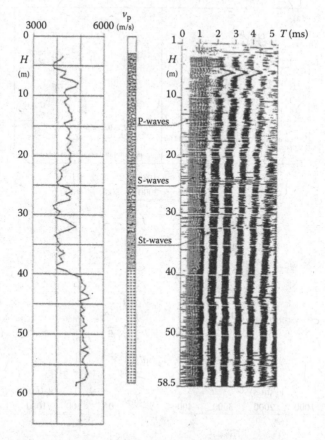

Figure 8.18 Measurement of the quality of jet grouting with sonic logging.

of a reduction of the injection pressure; in the lower part, from 2- to 8-m depth, the values of v_p are constantly high and similar with those found with laboratory tests, confirming the effectiveness of jet grouting in the coarse-grained soils; finally, the v_p values decrease rapidly at the bottom of the shaft, where soil composition changes, passing from coarse to clayey.

Cross-hole tests (ASTM D6760 2002) may be also used to check the effectiveness of jet grouting. As is well known, these tests have to be carried out with the source and the receiver placed at the same depth into two or more contiguous holes. Errors resulting from the imperfect verticality of the boreholes, which may turn into inaccurate estimates of the distance between the source and the receiver, must be minimised with the aid of inclinometric measurement of the position of holes. An example of an application of this technique is illustrated in Figure 8.19d and e, reporting

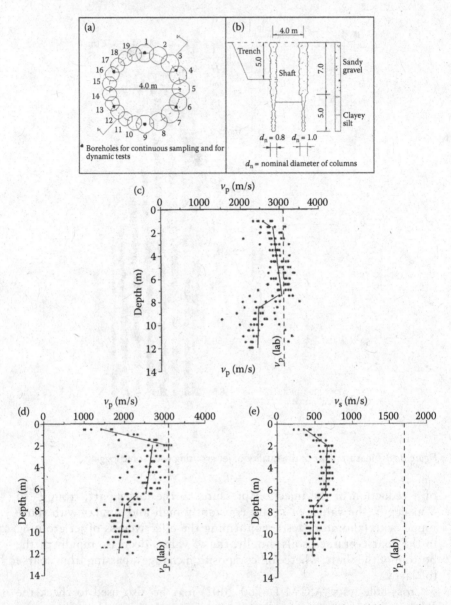

Figure 8.19 Various dynamic tests in a cylindrical jet-grouted shell (a and b): sonic logging (c); cross hole with compressive waves (d); cross hole with shear waves (e). lab, laboratory. (From Croce, P. et al., *Rivista Italiana di Geotecnica* 2: pp. 91–101 [in Italian], 1994.)

the trends of, respectively, compressive (v_p) and shear (v_s) wave velocities with depth, with reference to the case shown in Figure 8.19a and b. In this case, the cross-hole tests were carried out in the same holes previously used for the sonic logging tests. From a qualitative point of view, the results of cross-hole tests (in terms of both v_p and v_s) are similar with those obtained with sonic logging. However, the measured values of sonic (compressive) wave velocities and of cross-hole seismic compressive waves are much more scattered than the measured values of shear wave velocities, indicating that the latter are better suited to quantify the effects of soil improvement.

As previously shown, a tomography of compression or shear wave velocity gives very detailed information on the effects of jet grouting also in terms of continuity and mechanical properties. A good definition of the results can be nowadays obtained with multichannel acquisition seismographs, provided that sampling points are located at relatively close distances, and redundant information is obtained by inverting the position of the sources and the receivers.

An example of seismic tomography (using compression waves) of a rosette of seven partly overlapped columns is reported in Figure 8.20 (Ciufegni et al. 2007). The sampling points have been located along six boreholes (Figure 8.20a) and have been spanned at a vertical distance of 1 m (Figure 8.20b). The results show a good view of the parts having lower values of v_p and, therefore, the lower effectiveness of the treatment. However, the figure also shows that the shadowing fields do not overlap to the theoretically expected dimension of the columns, and geometrically unrealistic lateral extensions of jet grouting have been observed on the borders of the rosette. This is a possible effect of averaging calculation carried out with the inversion algorithm and can be minimised by refining the measuring grid.

Figure 8.21 reports the results of a tomography carried out using sonic waves, using the same data reported in Figure 8.19. Again, also in this example, tomography gives a complete view of the jet grouting effects.

The previously reported examples clearly show that the tomography represents, by far, the most accurate and complete *in situ* test to control the continuity and homogeneity of jet-grouted elements. However, it must be emphasised that investigation with this technique requires drilling several boreholes and adopting relatively sophisticated interpretation criteria that need noticeable expertise. As a consequence, tomography is, most of times, too expensive and time consuming to be implemented as a routine control of execution. However, it can be fruitfully adopted in the field trials or in the preliminary stages of construction to support or modify design choices.

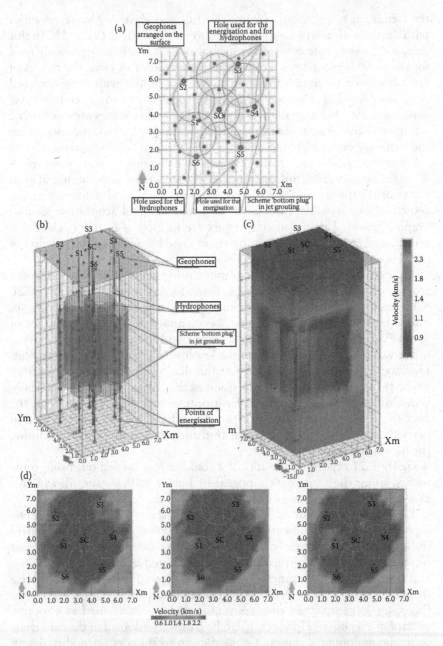

Figure 8.20 Tomography of cross-hole tests: (a) plan setup; (b) grid of measurement; (c) three-dimensional reconstruction of the shear wave velocities; (d) cross-section at three different depths (6, 8 and 10 m from the ground level). (From Ciufegni, S. et al., *Strade and Autostrade* 4-2007: pp. 156–162 [in Italian], 2007.)

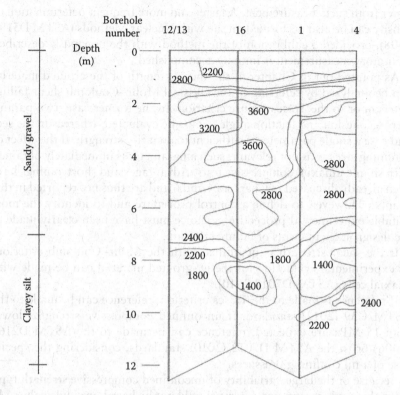

Figure 8.21 Contour map of P-wave velocities (metres per second) obtained from sonic tomography.

8.6.3 Physical and mechanical properties of the jet-grouted material

The physical and mechanical properties of jet-grouted materials are typically controlled by laboratory investigations performed using the same standards normally adopted for soils and rocks. Indirect control methods to be implemented on site after calibration with direct methods are also available to provide a faster and more frequent determination of properties.

Laboratory tests are generally carried out on cylindrical specimens cored from columns or from larger portions of jet-grouted materials. In special cases, cubic blocks are prepared by cutting samples of jet-grouted soil extracted from the walls of inspected trenches or shafts.

The dry unit weight is a very important physical property of the jet-grouted material and can be measured in the laboratory (ASTM D7263 2009) by coating samples with wax or by trimming specimens with a collar, wire saw or other suitable devices to give them uniform dimensions. Determination of water content (ASTM D2216 2010) is necessary to discount the role of pore water from such measurement. A faster and more frequent determination of density can be also performed on site with Nuclear Methods (ASTM D5195 2008), provided a calibration of the method with the previously described laboratory measurement is initially accomplished.

As pointed out in Chapter 4, the shear strength of the grouted material can be analysed by referring to the classical Mohr–Coulomb shear failure criterion or to the Tresca failure criterion. In the former case, two parameters – cohesion and friction angle – must be evaluated, whereas in the second case a single parameter is sufficient to quantify strength. If the effect of confining stresses is not relevant, such a parameter is immediately obtained with simple uniaxial compressive tests. Advantages and shortcomings have been already discussed in Chapters 4 and 6 and are thus not reported in this chapter. However, to define a control procedure and to identify the most suitable experimental procedure, a choice must have been clearly made at the design stage in terms of failure criterion.

If the shear strength is quantified with the Mohr–Coulomb criterion, its experimental evaluation for the jet-grouted material can be made with triaxial cells (ASTM D7012 2010).

If reference is made to the Tresca criterion, reference can be made to the ASTM C39 (2012) standard. If unconfined compressive strengths lower than 15 MPa are expected, reference can be made to the ASTM D2166 (2006) or to the ASTM D7012 (2010) standards, considering the special case of a nil confining pressures.

Because of the large variability of unconfined compressive strength typically observed, acceptance criteria should not be based on single values. As discussed in Chapter 4, a statistical approach can be conveniently used to interpret the experimental results.

Finally, laboratory permeability tests are sometimes carried on to check the waterproofing capability of jet-grouted materials, but it must be emphasised that results may lack of significance because the overall behaviour of the jet-grouted structure could be governed by discontinuities more than by the permeability of the treated material.

8.7 PERFORMANCE OF THE JET-GROUTED ELEMENTS

Performance tests allow a direct verification that jet-grouted elements meet the design requirements and, at the same time, represent an indirect checking of the correctness of predictions on the effects of treatment and on their regular execution. Basically, they consist of submitting samples of jet-grouted elements – singular columns or assemblies of columns, depending on the application to be tested – to the operative conditions and monitoring their response.

When jet grouting is used in conventional applications (e.g., as deep foundations, similarly with piles), the performance testing procedures can be simply taken from the traditional geotechnical experience. On the contrary, in the case of unusual applications, original testing equipment and procedures must be specifically conceived. In both cases, when the execution of tests on jet-grouted elements becomes difficult, if not impossible, the performance cannot be directly verified with experiments but must be indirectly assessed by a combination of mechanical assumptions, tests on the properties of jet-grouted elements and, if necessary, oversizing of treatments.

In the following section, a review of some typical investigation techniques is reported, along with prototype tests conceived for the peculiar applications of jet grouting.

8.7.1 Load tests

When jet grouting columns are used as isolated foundation reinforcements, loading tests can be carried out (Cicognani and Garassino 1989; Maertens and Maekelberg 2001; Bustamante 2002) to predict their actual performance and thus to confirm their adequacy for the support of the overlying structure. Axial load tests can be performed with the same procedures recommended for foundation piles (ASTM D1143/D1143M-e1 2007). Therefore, details on the testing procedure can be found in pile foundation books and manuals (e.g., Viggiani et al. 2011). Further test requirements, specifically of jet-grouted columns, usually consist of the regularisation of the column head (which must be plane and horizontal) and the insertion of a steel reinforcement into the upper part of the column to prevent local structural collapse induced by a possible misalignment of the testing load. In the case of pull-out tests, the reinforcement must be necessarily extended to the whole column.

Because of their dimensions and their irregular shape, jet-grouted columns are usually similar to large diameter piles, capable of transferring high loads to the surrounding soil (see Chapter 6). Therefore, the loading apparatus, the loaded members and the supporting frame must comply with the expected high loads.

An example of axial load test is reported in Figure 8.22, which shows the experimental setup and the results of compression and pull-out tests on four columns, all having a length of 7 m and a mean diameter of 0.90 m (Bzówka 2009). Three columns (P2, P3 and P4) were reinforced with HEB240 steel bars, whereas the other (P1) was left unreinforced (Figure 8.22a). The test load was applied by a hydraulic jack, and the reaction was provided by a set of steel beams anchored to the ground by steel-reinforced jet grouting columns of 11-m length (Figure 8.22b). Figure 8.22c shows the results of the compression tests on the unreinforced (P1) and the

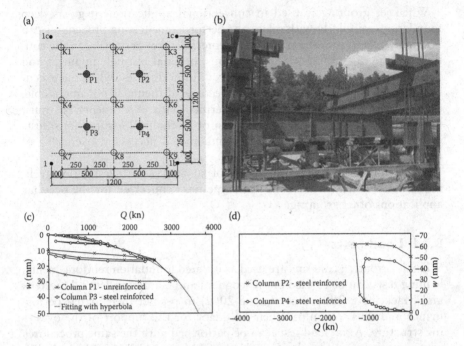

Figure 8.22 Axial loading tests on jet grouting columns (a) experimental setup;
(b) picture taken during a compression loading test; (c) results of compression test; (d) results of pull-out test. (From Bzówka, J., *Współpraca
kolumn wykonywanych techniką iniekcji strumieniowej z podłożem gruntowym
(Interaction of jet grouting columns with subsoil).* Monograph, Gliwice, Poland:
Silesian University of Technology Publishers [in Polish]: 244 p., 2009.)

steel-reinforced (P2) columns, in all qualitatively similar with that of piles.
The ultimate load can thus be computed by interpolating the experimental
data with a hyperbolic curve (Chin and Vail 1973) and extrapolating this
trend to the asymptotic value. In addition, the very close similarity of the
two curves shows that the two columns behave rather similarly, and therefore, no major defects influence the results that can be correlated to the
mean diameter of 0.90 m. Furthermore, reinforcement plays no role in the
compression test, as expected, considering that the load–settlement curve
is determined more by the deformation of the surrounding soil than by that
of the column. Figure 8.22d shows the results of two pull-out tests on the
reinforced columns P2 and P4. The sudden increase of displacements corresponds to the slippage between the steel reinforcement and the jet-grouted
material, and is a very useful experimental result to quantify the shear
strength of the steel–jet grouting interface.

Horizontal load tests can be also performed to check the performance
of jet grouting in applications, such as earth-retaining structures or

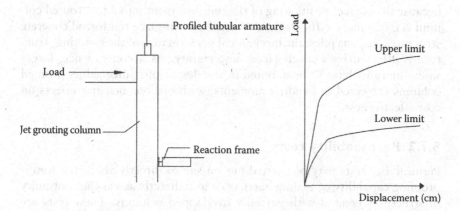

Figure 8.23 Transverse load tests on jet grouting columns. (Modified from Garassino A. L. and S. Palmoso, Prove di spinta orizzontale su colonne in terra stabilizzata. *Proceedings of the 15th National Geotechnical Conference 3*, Spoleto, Italy: pp. 154–161 [in Italian], May 4–6, 1983.)

horizontally loaded foundations, in which flexural stresses are relevant. Obviously, in this case, the tested columns must be reinforced because of the very limited resistance of the jet-grouted material to tensile stresses (which must be conservatively considered nil; see Chapter 6). If conventional horizontal load tests aimed to study the behaviour of the column immersed in soil have to be carried out, a horizontal load is applied at the column head, and the horizontal displacement is read.

An example of the setup and of the results of an unusual horizontal load test is reported in Figure 8.23 (Garassino and Palmoso 1983). In this example, the goal of the test was to evaluate the structural bending resistance of single columns, without any contribution of soil confinement. Therefore, the columns were excavated on one side, as shown in Figure 8.23a. A bending moment was applied by imposing with a jack a horizontal load to the column head and by inserting a horizontal reaction frame at the base of excavation, at a distance H from the column head.

Figure 8.23b shows that the horizontal load–displacement curves of a number of tests have a similar trend from a qualitative point of view but have a rather large scatter in the final result. The first part of the loading curve is, in all cases, stiffer, followed by a sudden increase of displacements and reduction of stiffness. The sharp change in mechanical behaviour corresponds to the formation of a plastic hinge in the column, with the yielding of the internal steel reinforcement. The large scatter of the limit value of the horizontal load mostly depends on the position of the reinforcing bar within each column, randomly deviating from the centreline. This experimental evidence, which has been confirmed by similar results from the authors' personal experience, is of paramount importance:

because the correct positioning of the reinforcement into a jet-grouted column is much more difficult than in usual cast-in-place reinforced concrete structures, such as piles, the mechanical performance of the resulting structure is affected by a much larger uncertainty. As a consequence, larger safety margins have to be adopted in the design of reinforced jet-grouted columns subjected to bending moments, with obvious negative effects on cost effectiveness.

8.7.2 Permeability tests

Permeability tests may be carried out on site to directly check the water-proofing capability of sealing barriers or to indirectly assess the continuity of structures created with partially overlapped columns. These tests are performed by drilling boreholes in the jet-grouted body and by establishing a piezometric head difference between the inside of the hole and the surrounding mass (ASTM D5084 2010). Measurements can be extended throughout the hole to detect a larger portion of jet-grouted material with a single test or may be repetitively performed at different depths, creating smaller filtering sections confined by a couple of packers.

Considering the typically low permeability of well-treated jet-grouted materials and the very long time required to establish steady-state flow conditions, falling head infiltration tests are typically performed. In these transient state tests, the permeability coefficient can be related to the lowering of water level in the hole, recorded as a function of time (ASTM D4750 2001). If the deformability of the confining packers is not negligible, it should be considered in the interpretation of the experimental results.

Since the saturation degree may strongly affect the results of these pumping tests, other testing methods typical of rock masses can be used (Ahmed et al. 1991). For instance, the calculation of permeability can be attempted testing a larger portion of the jet-grouted mass through the so-called 'multiwell interference test': it consists of applying transient piezometric conditions in one well and measuring variations in the other wells located in the surrounding portion of the material to check if the wells are hydraulically connected (i.e., the jet-grouted mass has discontinuities).

An example of such measurement is reported in Figure 8.24. The test was performed to prove the potential and show the possible limitations of the jet grouting technique to waterproof tunnel excavation in the city of Barcelona (Arroyo et al. 2012). An almost full-scale (80%) heading section of the tunnel was prepared at a true excavation depth of 15 m (Figure 8.24a). The trial tunnel heading section was created below the groundwater table, at the bottom part in a clayey formation and at the top part in a sandy layer, which extended for 9 m above its crown. The canopy was conceived to isolate the tunnel from the surrounding water in the permeable sandy soil. The test tunnel contour canopy consisted of two concentric

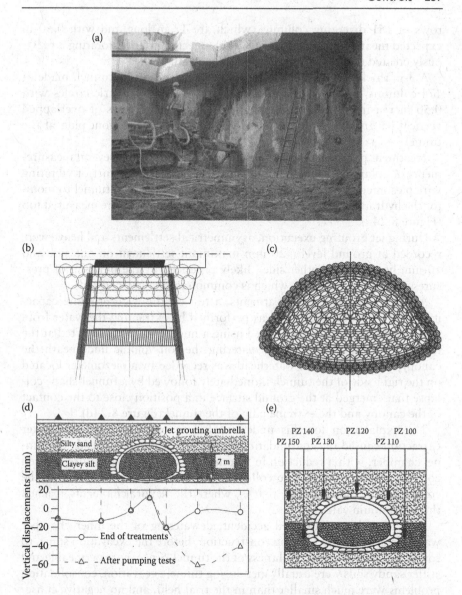

Figure 8.24 Trial field for tunnel execution (a) picture of the front; (b) horizontal section; (c) cross-section; (d) profile of settlements at ground level induced by the execution of treatments and the extraction of water from the tunnel chamber; (e) cross-section with position of piezometers. (From Arroyo, M. et al., Design of jet grouting for tunnel waterproofing, In Viggiani, ed., *Proceedings of the 7th International Symposium on the Geotechnical Aspects of Underground Construction in Soft Ground*: TC28-IS Rome: London, United Kingdom: Taylor & Francis Group: pp. 181–188, 2012.)

rows of 151 diverging columns, which are 12 m long and with 0.50 m expected mean diameters (Figure 8.24b), injected after perforating a previously created thick jet grouting supporting wall.

A 3-m thick plug was created at the deeper end of the tunnel, made of 174 columns of 0.80 m diameter, located on six concentric circles with 0.50 increasingly larger radii (Figure 8.24c). Three rows of overlapped vertical jet grouting columns were created to form the front plug of the tunnel.

Treatment parameters were continuously recorded, and several measurements of column axis inclination were also performed. A net of vibrating wire piezometers was installed above and around the trial tunnel to monitor the hydraulic heads, and ground surface settlements were measured too (Figure 8.24d).

During jet grouting execution, asymmetrical settlements and heave were recorded at ground level (~20-mm maximum settlement on one side and 10-mm heave on the other side), likely produced by the build-up of pressures induced by injection, which is common in fine-grained soils.

After the conclusion of treatments, a test on the waterproofing capacity of the jet-grouted canopy was performed by extracting the water from the tunnel volume to be excavated using a number of pipes inserted at the tunnel front. Few days after dewatering the soil volume underneath the canopy, a sudden drop of waterhead was recorded by a piezometer located on the right side of the tunnel, immediately followed by a funnel-shape collapse that emerged at the ground surface in a position close to the contact of the canopy and the extreme plug of the tunnel (Figure 8.24d).

The explanation for this undesired failure mechanism is that water extraction caused the consolidation of the fine-grained soil inside the tunnel chamber, with a reduction in volume and bending of the unreinforced columns of the canopy. The collapse occurred at the intersection of the columns with the end plug, that is, where the flexural moments reached their maximum values.

Despite such an undesired accident, dewatering of the inner chamber was satisfactorily used during construction, before the excavation stage, to routinely check the water tightness of the tunnel. Since more permeable and stiffer sandy soils were usually met during tunnel excavation, consolidation problems were much smaller than in the trial field, and no negative consequences of water extraction were detected.

8.8 MONITORING OF THE SURROUNDING STRUCTURES

The monitoring of the surrounding environment is conceived to check in field trials or to control during construction if particularly sensitive structures or environments may be negatively affected by jet grouting operations. The strategy of these controls is to highlight potentially negative phenomena from their very early stage, to activate appropriate countermeasures before critical consequences may occur, and therefore, promptness of measurement and quick data processing are fundamental requirements of the monitoring system. These goals can be, nowadays, pursued with automated recording instruments communicating with computers via data loggers. Dedicated software can store the data and trigger alarm messages when predefined threshold values of the critically monitored parameters are exceeded.

The most critical issues from an environmental point of view are the diffusion of contaminants, noise and vibrations. The first problem can be faced by measuring the concentration of pollutants on samples periodically taken from water-bearing strata or from watercourses at some distance from the injection point. An excessive presence may indicate that an uncontrolled flow of the injected mixtures is occurring underground or that an inaccurate collection of the spoil at the borehole head is occurring.

Noises and vibrations can be measured very efficiently by a network of geophones strategically positioned in the surroundings. Once it has been concluded that these effects are not determined by other environmental factors, such as traffic or other machinery working nearby, they can be reduced by replacing the drilling and injecting tools with more efficient ones or by modifying the injection procedure.

Heave and/or settlement of the nearby ground surface are probably the most frequent disturbances that may occur when intensive jet grouting treatments are performed. They can be normally disregarded when injections are executed far from buildings, infrastructures or lifelines, but consequences may be particularly severe when jet grouting is performed near structures sensitive to the ground deformation, as it is typical in urban environments.

Accurate real-time monitoring of these phenomena can be implemented, thanks to a variety of topographic instruments that may be set to automatically perform readings with prescribed time intervals. Vertical movements with reference to absolute reference systems can be obtained by GPS aerials or by settlement cells hydraulically connected to a fixed reservoir; differential settlement profiles can be traced along prescribed alignments by a sequence of electronic liquid-level gauges or subhorizontal inclinometers placed in the ground; tilting of buildings can be measured with wall inclinometers, whereas deformation and cracks can be read, respectively, by extensometers or displacement transducers.

In the most advanced systems, a complete pattern of movements can be rapidly be obtained on wide areas by installing a total station set to perform periodical measurements on optical reflector targets positioned all around. The kinematic monitoring of the ground surface can be also supported by underground observations, including measurement of pore-water pressure by multilevel vibrating wire piezometers, longitudinal strains by single or multipoint rod extensometers and transversal deformation by in-place inclinometers.

Two different examples of such global monitoring systems are described in the following to highlight the possible opposite effects of jet grouting. The first case (Russo and Modoni 2005; Croce et al. 2004b) refers to the excavation of a tunnel underpassing an important road near the city of Florence (Italy) (Figure 8.25a). The designed construction procedure prescribed a full-face excavation of the subcircular cross section (11 m high

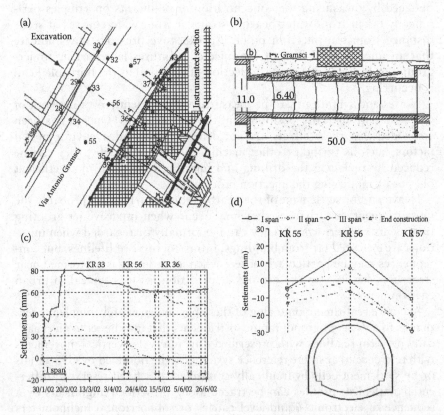

Figure 8.25 Monitoring of the effects on the surrounding of a tunnel: (a) plan view; (b) longitudinal section; (c) time sequence of settlements on three surveys at ground level; (d) ground settlement profiles at a cross-section. (From Russo, G. and G. Modoni, Monitoring results of a tunnel excavation in urban area. *Proceedings of the 5th International Symposium of the Technical Committee TC28: Geotechnical Aspects of Underground Construction in Soft Ground,* Amsterdam, Netherlands, June 15–17, 2005: pp. 751–756, 2005.)

and 15 m wide) and a subdivision of the tunnel length (~50 m) into seven sectors of equal length (6.4 m). The front face was reinforced with grout injected from 24-m–long fibreglass tubes; canopies formed by 71 jet grouting columns of 0.5-m mean diameter and 12-m length were created to support the tunnel contour (Figure 8.25b).

The area above the tunnel to be realised was monitored by a considerable number of instruments (Figure 8.25a), including a high-precision optical levelling located on a grid of 35 benchmarks at the ground level, electronic liquid-level gauges installed along two transversal sections at limited depth from the ground level and two multilevel extensometers located in the ground near the instrumented sections. Deformation of the ground was also indirectly monitored by five clinometers installed on the walls of neighbouring buildings and by a vertical inclinometer placed near one instrumented section, whereas pore-water pressure was measured by two Casagrande piezometers placed at different depths on an instrumented section.

Figure 8.25c shows the time sequence of settlements measured on the three surveys located along the centreline of the tunnel (KR33, KR56 and KR36 in the figure) during the execution of jet grouting in each sector, marked with the vertical dashed lines. The jet grouting performed in the first span of the tunnel produced a substantial lift at ground level, with approximately 80 mm of upward movement recorded at point KR33, 30 mm at point KR56 and almost nil movement at point KR36. Considering the clogging of boreholes and the lack of spoil responsible for this effect, the jet grouting procedure was modified, and a double passage of the monitor was performed into the hole with increased retraction rates and reduced injection flow rates. With such a change in column production, the uplift produced by treatments in the subsequent spans could be reduced. As shown in Figure 8.25c, uplift was subsequently compensated by downward settlements that occurred after excavation. A complete pattern of movements can be seen in Figure 8.25d, in which the profiles taken at different times in the cross-section passing through points KR55, KR56 and KR57 are traced.

The second example concerns the construction of a train station in the city of Barcelona (Eramo et al. 2012). A plug was preliminarily made with jet grouting to seal the bottom of the excavation, having plan dimensions of 110 m for 30 m and a thickness of 10 m (Figure 8.26a, b). Since the station located in a densely populated area and surrounded by relatively tall buildings, a monitoring plan was conceived to measure settlements in the surrounding area during injection and excavation. To this aim, a high-precision optical levelling was carried out with high frequency on a grid of 35 benchmarks located around the station (Figure 8.26c).

The interpolation of these data with a geostatistical method gives, at different times, settlement maps such as the two reported in Figure 8.26c, describing the situation before and after the execution of jet grouting. Their

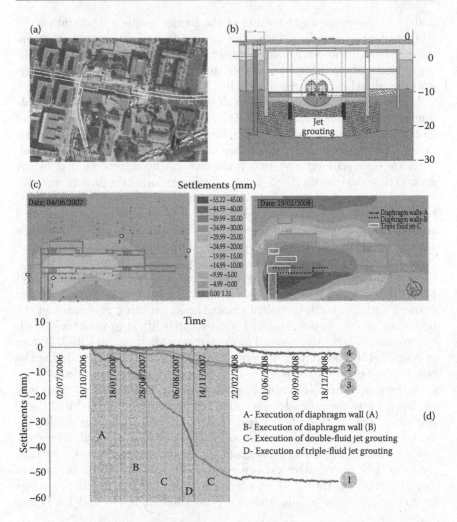

Figure 8.26 Monitoring of the effects on the surrounding of a bottom plug: (a) plan view; (b) typical cross-section; (c) contour map of settlements; (d) settlement time history on four monitored positions. (From Eramo, N. et al., Design control and monitoring of a jet-grouted excavation bottom plug, In G. Viggiani, ed., *Proceedings of the 7th International Symposium on the Geotechnical Aspects of Underground Construction in Soft Ground*: London: Taylor & Francis Group: pp. 611–618, 2012.)

direct comparison shows that a significant downward settlement was produced by jet grouting operations. In this site, similar evidence was already obtained during some field trials conducted prior to construction, when triplets of columns were created with different injection systems and settlements measured at different distances. Although not clearly demonstrated, these effects may be related to the consolidation phenomena induced by the retarded hardening of cement and by the heat developed during setting reactions.

From the two settlement maps reported in Figure 8.26c, it is also seen that the most subsiding area, with the largest settlements of approximately 55 mm, is located near the ramp built to access the construction site, where a diaphragm wall was realised. However, the time evolution on four different benchmarks (Figure 8.26d) shows that less than 20 mm were induced by the construction of diaphragm walls, whereas the remaining part (~35 mm) occurred during and after the execution of jet grouting treatments. It is also worth observing that injections with a triple-fluid system caused a larger amount of settlements. This method, initially attempted to speed up operations, was then abandoned after observing the sudden increase of settlement rates on closer benchmarks.

References

AGI. 2012. *Jet Grouting Guidelines*: Associazione Geotecnica Italiana: 69 p [in Italian].

Ahmed, U., S. F. Crary, and G. R. Coates. 1991. Permeability estimation: The various sources and their interrelationships. *Journal of Petroleum Technology*, May 1991: pp. 578–587.

Alzamora, D., M. H. Wayne, and J. Han. 2000. Performance of SRW supported by geogrids and jet-grout columns. *Proceedings of ASCE Specialty Conference on Performance Confirmation of Constructed Geotechnical Facilities*: Geotechnical Special Publication 94: pp. 456–466.

Andrus, R. D. and R. M. Chung. 1995. *Ground Improvement Techniques for Liquefaction Remediation Near Existing Lifelines*, Gaithersburg, MD: U.S. Department of Commerce: 82 p.

API. 2009. *Recommended Practice for Field Testing Water-Based Drilling Fluids*, API RP 13B-1, 4th ed.: Washington, DC: American Petroleum Institute: 91 p.

Arroyo, M., A. Gens, E. Alonso, G. Modoni, and P. Croce. 2007. *Informes Sobre Tratamientos de Jet Grouting*. ADIF LAV Madrid-Barcelona-Francia, Tramo Torrasa-Sants. Report of the Universidad Politecnicha de Catalunya: 110 p [in Spanish].

Arroyo, M., A. Gens, P. Croce, and G. Modoni. 2012. Design of jet grouting for tunnel waterproofing, In G. Viggiani, ed., *Proceedings of the 7th International Symposium on the Geotechnical Aspects of Underground Construction in Soft Ground*: TC28-IS Rome: London, United Kingdom: Taylor & Francis Group: pp. 181–188.

Aschieri, F., M. Jamiolkowski, and R. Tornaghi. 1983. Case history of cutoff wall executed by jet grouting. *Proceedings of the 8th ESCMFE* 1: pp. 121–126.

ASTM. 1995. *Standard Guide for Planning and Conducting Borehole Geophysical Logging*, Designation D5753-95e1: West Conshohocken, PA: American Society for Testing and Materials: 9 p.

ASTM. 1999. *Standard Practice for Evaluation of Laboratories Testing Hydraulic Cement*, Designation C1222-99: West Conshohocken, PA: American Society for Testing and Materials: 4 p.

ASTM. 2001. *Standard Test Method for Determining Subsurface Liquid Levels in a Borehole or Monitoring Well (Observation Well)*, Designation D4750-87: West Conshohocken, PA: American Society for Testing and Materials: 6 p.

ASTM. 2001. *Standard Test Method for Time of Setting of Hydraulic Cement by Vicat Needle*, Designation C191-01a: West Conshohocken, PA: American Society for Testing and Materials: 8 p.

ASTM. 2002. *Standard Specification for Portland Cement*, Designation C150-02: West Conshohocken, PA: American Society for Testing and Materials: 9 p.

ASTM. 2002. *Standard Test Method for Integrity Testing of Concrete Deep Foundations by Ultrasonic Cross-Hole Testing*, Designation D6760-02: West Conshohocken, PA: American Society for Testing and Materials: 7 p.

ASTM. 2006. *Standard Test Method for Unconfined Compressive Strength of Cohesive Soil*, Designation D2166-06: West Conshohocken, PA: American Society for Testing and Materials: 6 p.

ASTM. 2007. *Standard Test Methods for Deep Foundations Under Static Axial Compressive Load*, Designation D1143/D1143M-07e1: West Conshohocken, PA: American Society for Testing and Materials: 15 p.

ASTM. 2008. *Standard Test Method for Density of Soil and Rock in Place at Depths Below Surface by Nuclear Methods*, Designation D5195-08: West Conshohocken, PA: American Society for Testing and Materials: 5 p.

ASTM. 2008. *Standard Test Method for Determining Rock Quality Designation (RQD) of Rock Core*, Designation D6032-08: West Conshohocken, PA: American Society for Testing and Materials: 5 p.

ASTM. 2009. *Standard Test Method for Marsh Funnel Viscosity of Clay Construction Slurries*, Designation D6910/D6910M-09: West Conshohocken, PA: American Society for Testing and Materials: 3 p.

ASTM. 2009. *Standard Test Methods for Laboratory Determination of Density (Unit Weight) of Soil Specimens*, Designation D7263-09: West Conshohocken, PA: American Society for Testing and Materials: 7 p.

ASTM. 2010. *Standard Test Method for Compressive Strength and Elastic Moduli of Intact Rock Core Specimens Under Varying States of Stress and Temperatures*, Designation D7012-10: West Conshohocken, PA: American Society for Testing and Materials: 9 p.

ASTM. 2010. *Standard Test Method for Expansion and Bleeding of Freshly Mixed Grouts for Preplaced-Aggregate Concrete in the Laboratory*, Designation C940-10a: West Conshohocken, PA: American Society for Testing and Materials: 3 p.

ASTM. 2010. *Standard Test Methods for Laboratory Determination of Water (Moisture) Content of Soil and Rock by Mass*, Designation D2216-10: West Conshohocken, PA: American Society for Testing and Materials: 7 p.

ASTM. 2010. *Standard Test Methods for Measurement of Hydraulic Conductivity of Saturated Porous Materials Using a Flexible Wall Permeameter*, Designation D5084-10: West Conshohocken, PA: American Society for Testing and Materials: 23 p.

ASTM. 2012. *Standard Guide for Field Logging of Subsurface Explorations of Soil and Rock*, Designation D5434-12: West Conshohocken, PA, American Society for Testing and Materials: 3 p.

ASTM. 2012. *Standard Guide for Planning and Conducting Borehole Geophysical Logging*, Designation D5753-05: West Conshohocken, PA: American Society for Testing and Materials: 9 p.

ASTM. 2012. *Standard Specification for Coal Fly Ash and Raw or Calcined Natural Pozzolan for Use in Concrete*, Designation C618-12a: West Conshohocken, PA: American Society for Testing and Materials: 5 p.

ASTM. 2012. *Standard Specification for Mixing Water Used in the Production of Hydraulic Cement Concrete*, Designation C1602/C1602M-12: West Conshohocken, PA: American Society for Testing and Materials: 5 p.

ASTM. 2012. *Standard Test Method for Compressive Strength of Cylindrical Concrete Specimens*, Designation C39/C39M-12a: West Conshohocken, PA: American Society for Testing and Materials: 7 p.

ASTM. 2012. *Standard Test Method for Compressive Strength of Hydraulic Cement Mortars (Using 2-in or 50-mm Cube Specimens)*, Designation C109/C109M-12: West Conshohocken, PA: American Society for Testing and Materials: 10 p.

ASTM. 2013. *Standard Specification for Concrete Aggregates*, Designation C33/C33M-13: West Conshohocken, PA: American Society for Testing and Materials: 11 p.

Attewill, L. J. S., J. D. Gosden, D. A. Bruggemann, and G. C. Euinton. 1992. The construction of a cutoff in a volcanic residual soil using jet grouting, In A. Charles, N. M. Parr, and S. Walker, eds., *Water Resources and Reservoir Engineering*: London, United Kingdom: Thomas Telford: pp. 131–138.

Baecher, G. B. and J. T. Christian. 2003. *Reliability and Statistics in Geotechnical Engineering*: John Wiley & Sons, Ltd.: 605 p.

Balossi Restelli, A., A. Colombo, F. Gervaso, and P. Lunardi. 1986. Tecnologie speciali per il preconsolidamento di scavi nelle alluvioni di Milano in occasione della costruzione della linea 3 della Metropolitana, In *Proceedings of the International Conference 'Grandi Opere Sotterranee'*, Florence, Italy, June 8–11, 1986: pp. 612–618 [in Italian].

Balossi Restelli A. and M. Profeta. 1985. Autostrada Udine-Tarvisio. Trattamenti speciali per fondazioni e difese fluviali. *Industria delle Costruzioni*, December: pp. 29–42 [in Italian].

Bell, A. L. 1983. *Control of Column Geometry: Jet Grouting in Sand*: GKN Keller Technical Report: 77 p.

Bell, A. L. 1993. Jet grouting, In M. P. Moseley, ed., *Ground Improvement*: Boca Raton, FL: Blackie: pp. 149–174.

Berg, J. and H. Samol. 1986. Soilcrete: A method of soil improvement applied in soil and water engineering project. *Wasser und Boden* 9: pp. 451–455.

Bergschneider, B. 2002. *Zur Reichtweite beim Düsenstrahlverfahren im Sand*: PhD thesis, Bergische Universität Wuppertal Fachbereich Bauingenieurwesen: 176 p [in German].

Bergschneider, B. and B. Walz. 2003. Jet grouting: Range of the grouting jet, In Vanicek et al., eds., *Proceedings of the 13th European Conference on Soil Mechanics and Foundation Engineering*: Prague, Czech Republic: Czech Geotechnical Society: pp. 53–56.

Botto, G. 1985. Developments in the techniques of jet grouting, *12th Ciclo di Conferenze di Geotecnica*: Torino, reprint by Trevi: pp. 81–90.

Botto, G. and F. Capolupo. 1989. Trattamenti colonnari di gettiniezione normale (jet grout monofluido) in formazioni argillose per le fondazioni della centrale gas dell'Agip di Falconara: Ricerca della tecnologia ottimale a mezzo di campo prove. *Proceedings of the 12th Nationali Geotechnical Conference*, Taormina, Italy, April 26–28, 1989, Associazione Geotecnica Italiana: pp. 47–53 [in Italian].

Brinch Hansen, J. B. 1970. A revised and extended formula for bearing capacity. *Bulletin No. 28*: Danish Geotechnical Institute: pp. 3–21.

British National Society of ISSMFE. 1963. *Proceedings of the Symposium on Grouts and Drilling Muds in Engineering Practice*: London, United Kingdom: Butterworths: pp. 63–67.

Broms, B. B. 1964a. Lateral resistance in cohesive soils. *Journal of Soil Mechanics and Foundations Division* 90(SM2): pp. 27–63.

Broms, B. B. 1964b. Lateral resistance in cohesionless soils. *Journal of Soil Mechanics and Foundations Division* 90(SM3): pp. 123–156.

Brown, E. T. and M. V. Barr. 1978. 'Instrumented Drilling as an Aid to Site Investigations', *Proc. 3rd Intern. Proc. of the III Conf. of the International Association of Engineering Geology*, Madrid, Section IV, Vol. 1, pp. 21–28.

BS 8500-1. 2006. Concrete. *Complementary British Standard to BS EN 206-1: Part 1. Method of Specifying and Guidance for the Specifier*: Greater London, United Kingdom: British Standard Institution: 42 p.

Burke, G. K. 2007a. Vertical and horizontal groundwater barriers using jet-grout panels and columns (GSP 168). *Proceedings of Geo-Institute's Geo-Denver: New Peaks in Geotechnics*, Denver, CO, February 18–21, 2007: 10 p.

Burke, G. K. 2007b. New methods for underpinning and earth retention, grouting for ground improvement (GSP 168). *Proceedings of Geo-Institute's Geo-Denver: New Peaks in Geotechnics*, Denver, CO, February 18–21, 2007: 12 p.

Burke, G. K. and H. Yoshida. 2013. Jet grouting, In K. Kirsch and A. Bell, eds., *Ground Improvement*, 3rd ed.: CRC Press: Taylor & Francis Group: pp. 207–258.

Bustamante, M. 2002. Les colonnes de jet grouting. *Report of the Seminar: Pathologies des Sols et des Foundations*, http://www.keller-france.com/recherche-et-developpement/theses-et-publications: 6 p [in French].

Bzówka, J. 2009. *Współpraca kolumn wykonywanych techniką iniekcji strumieniowej z podłożem gruntowym (Interaction of jet grouting columns with subsoil)*. Monograph, Gliwice, Poland: Silesian University of Technology Publishers: 244 p [in Polish].

Chen, W. F. 1975. *Limit Analysis and Soil Plasticity*. Amsterdam, Netherlands: Elsevier: 638 p.

Chin, F. K. and A. J. Vail. 1973. Behavior of piles in alluvium. *Proceedings of the 8th International Conference on Soil Mechanics and Foundation Engineering* 2.1, Moscow, Russia, July 4–6: pp. 47–52.

Chu, E. H. 2005. *Turbulent Fluid Jet Excavation in Cohesive Soil, With Particular Application to Jet Grouting*: PhD thesis, Massachusetts Institute of Technology, Cambridge, MA: 456 p.

Cicognani, M. and A. L. Garassino. 1989. Controlli nell'esecuzione dei trattamenti con jet grouting. *Proceedings of the 17th National Geotechnical Conference 1*, Associazione Geotecnica Italiana: pp. 97–105 [in Italian].

Ciufegni, S., G. Marcheselli, S. Sdoga, L. Bellicchi, and G. Bruzzi. 2007. Il nuovo ponte sul fiume po a ostiglia. *Strade and Autostrade* 4-2007: pp. 156–162 [in Italian].

Clough, G. W., M. Kuckw, and G. Kasali. 1979. Silicate stabilized sands. *ASCE Journal of the Geotechnical Engineering Division* 105: pp. 65–82.

Conn, A. F. and M. T. Gracey. 1988. Using water jets to wash out explosives and propellants. *Proceedings of the 9th International Symposium on Jet Cutting Technologies*, Paper F4: pp. 242–250.

Coomber, D. B. 1985a. Groundwater control by jet grouting. *Proceedings of the Annual Conference of the Engineering Group of the Geological Society*, Sheffield, England, September 15–19, 1985, The Geological Society of London: pp. 445–454.

Coomber, D. B. 1985b. Tunelling and soil stabilisation by jet grouting. *Proceedings of the 4th International Symposium on Tunelling*, Brighton, England, June 1985: Paper 22: pp. 277–283.

Coomber, D. B. and D. W. Wright. 1984. Jet grouting at Felixstowe docks. *Ground Engineering*: 15(2): pp. 32–36.

Costabile, M. V., A. Flora, G. P. Lignola, and G. Manfredi. 2005. A semiprobabilistic approach to the design of jet-grouted umbrellas in tunelling, In M. I. M. Pinto and I. Jefferson, eds., *Proceedings of the 6th International Conference on Ground Improvement Techniques*. Coimbra, Portugal, July 18–19, 2005: pp. 261–272.

Covil, C. S. and A. E. Skinner. 1994. Jet grouting: A review of some of the operating parameters that form the basis of the jet grouting process, In *Grouting in the Ground*: London, United Kingdom: Thomas Telford: pp. 605–627.

Croce, P. and A. Flora. 1998. Effects of jet grouting in pyroclastic soils. *Rivista Italiana di Geotecnica* 2: pp. 5–14.

Croce, P. and A. Flora. 2001. Analysis of single fluid jet-grouting. Closure to discussion. *Geotechnique* 51(10): pp. 905–906.

Croce, P. and A. Flora. 2006. Analysis of single-fluid jet grouting. *Geotechnique* 50(6): pp. 739–748.

Croce, P. and G. Modoni. 2006. Design of jet grouting cutoffs. *Ground Improvement* 10(1): pp. 1–9.

Croce, P. and G. Modoni. 2008. Design of jet grouting cutoffs: Closure. *Ground Improvement* 13(1): pp. 47–50.

Croce, P. and G. Modoni. 2010. Consolidamento delle fondazioni di rilevati stradali e ferroviari. *Rivista Italiana di Geotecnica* 44(3): pp. 30–45 [in Italian].

Croce, P., A. Chisari, and T. Merletti. 1990. Indagini sui trattamenti dei terreni mediante jet grouting per le fondazioni di alcuni viadotti autostradali. *Rassegna dei Lavori Pubblici*, November: pp. 249–260 [in Italian].

Croce, P., S. di Maio, G. Speciale, and L. Cassibba. 2012a. Design and construction of sewer tunnel in difficult site conditions, In G. Viggiani, ed., *Proceedings of the 7th International Symposium on the Geotechnical Aspects of Underground Construction in Soft Ground*: TC28-IS Rome: London, United Kingdom: Taylor & Francis Group, May 2011: pp. 157–164.

Croce, P., A. Flora, S. Lirer, and G. Modoni. 2012b. Prediction of jet grouting efficiency and column average diameter. *Proceedings of the ICSMGE–TC211 International Symposium on Ground Improvement IS-GI 4*, Brussels, Belgium, May 31–June 1, 2012: pp. 215–224.

Croce, P., A. Flora, and G. Modoni. 2001. Experimental investigation of jet grouting. *Proceedings of the ASCE Conference '2001 a GeoOdissey'*, Blacksburg, VA: Virginia Tech, June 9–13: pp. 245–259.

Croce, P., A. Flora, and G. Modoni. 2004a. *Jet Grouting: Tecnica, Progetto e Controllo*. Benevento, Italy: 221 p [in Italian].

Croce, P., A. Gajo, L. Mongiovì, and A. Zaninetti. 1994. Una verifica sperimentale degli effetti della gettiniezione. *Rivista Italiana di Geotecnica* 2: pp. 91–101 [in Italian].

Croce, P., G. Modoni, and M. F. W. Carletto. 2011. Correlazioni per la previsione del diametro delle colonne di jet grouting. *Proc. of the XXIV National Geotechnical Conference, 'Innovazione Tecnologica nell'Ingegneria Geotecnica'*, Napoli, Italy, June 22–24, 2011, Associazione Geotecnica Italiana: pp. 423–430 [in Italian].

Croce, P., G. Modoni, and F. Palmisano. 2006. Progetto di fondazioni a pozzo realizzate con la tecnica del jet grouting. *Proceedings of the National Conference of the Researchers of Geotechnical Engineering*, Bari, Italy, September 3–4: pp. 327–340 [in Italian].

Croce, P., G. Modoni, and G. Russo. 2004b. Jet grouting performance in tunnelling. *Proceedings of the Conference on Deep Foundations, Earth-Retaining Structures, Soil Improvement, and Grouting—Geo-Support 2004*, Orlando, FL, September 2003: pp. 910–922.

Dabbagh, A. A., A. S. Gonzalez, and A. S. Peña. 2002. Soil erosion by a continuous water jet. *Soils and Foundations* 42(5): pp. 1–13.

Davidson, P. A. 2004. *Turbulence*: Oxford, United Kingdom: Oxford University Press: 678 p.

Davie, J., M. Pyial, A. Anver, and B. Tekinturhan. 2003. Jet-grout columns partially support natural draft cooling tower. *Proceedings of the 12th Pan-American Conference on Soil Mechanics and Geotechnical Engineering*: Cambridge, MA: ASCE GeoInstitute: pp. 365–376.

de Paoli, B., R. Tornaghi, and D. A. Bruce. 1989. Jet grout stabilization of peaty soils under a railway embankment in Italy. *Foundation Engineering: Current Principles and Practice*, New York: ASCE: pp. 272–290.

de Vleeshauwer, H. and G. Maertens. 2000. Jet grouting: State of the art in Belgium. *Proceedings of the Conference 'Grouting: Soil Improvement—Geosystem Including Reinforcement'*: Helsinki, Finland: Finnish Geotechnical Society: pp. 145–156.

di Natale, M. and R. Greco. 2000. Misure di velocità in un getto sommerso ad asse verticale mediante la tecnica PIV. *Proceedings of the Symposium Promoted by Associazione Italiana di Anemometria Laser (AIVELA)*, Ancona, Italy, October 2000: 11 p [in Italian].

DIN 4093. 2012. Design of ground improvement: Jet grouting, deep mixing, or grouting. *Standard of the Deutsches Institut für Normung*, 2012 ed.: Düsseldorf, Germany: 17 p.

Durgunoğlu, H. T., H. F. Kulaç, K. Oruç, R. Yıldız, J. Sickling, I. E. Boys, T. Altugu, and C. Emrem. 2003. A case history of ground treatment with jet grouting against liquefaction for a cigarette factory in Turkey. *Grouting and Grout Treatment*, February 2003: pp. 452–463.

Durgunoglu, T., H. F. Kulaç, S. Ikiz, O. Sevim, and O. Akcakal. 2012. Offshore jet grouting: A case study. *Proceedings of the ISSMGE–TC211 International Symposium on Ground Improvement*, Brussels, Belgium, May 31–June 1, 2012: pp. 225–234.

EN 12716. 2001. *Execution of Special Geotechnical Works: Jet Grouting*. European Committee for Standardization, 35 p.

EN 1997-1. 2004. *Eurocode 7: Geotechnical Design*. European Committee for Standardization: 170 p.

ENEL. 2000. Forcoletta Dam (Italy): Diaframma di tenuta a monte in jet grouting. Ente Nazionale per l'Energia Elettrica, Internal Report: 49 p [in Italian].

Eramo, N., G. Modoni, and M. Arroyo. 2012. Design control and monitoring of a jet-grouted excavation bottom plug, In G. Viggiani, ed., *Proceedings of the 7th International Symposium on the Geotechnical Aspects of Underground Construction in Soft Ground*: London: Taylor & Francis Group: pp. 611–618.

Falcão, J., A. Pinto, and F. Pinto. 2001. Case histories and work performance of vertical jet grouting solutions. *Proceedings of the 4th International Conference on Ground Improvement*: Helsinki (Finland): Finnish Geotechnical Society: pp. 165–171.

Fang, Y. S. and Y. C. Chung. 1997. Jet grouting for shield tunelling in Taipei. *Proceedings of the International Conference on Ground Improvement Techniques*, Macau, People's Republic of China, May 6–8, 1997: pp. 189–196.

Fang, Y. S., L. Y. Kuo, and D. R. Wang. 2004. Properties of soilcrete stabilized with jet grouting. *Proceedings of the 14th International Offshore and Polar Engineering Conference*, Toulon, France, May 23–28, 2004: pp. 696–702.

Fang, Y. S., J. J. Liao, and T. K. Lin. 1994a. Mechanical properties of jet-grouted soilcrete. *Quarterly Journal of Engineering Geology and Hydrogeology* 27: pp. 257–265.

Fang, Y. S., J. J. Liao, and S. C. Sze. 1994b. An empirical strength criterion for jet-grouted soilcrete. *Engineering Geology* 37: pp. 285–293.

Farmer, W. and P. B. Attewell. 1965. Rock penetration by high-velocity jet. *International Journal of Rock Mechanics and Mining Sciences* 2: pp. 135–153.

Fenton, G. A. and D. V. Griffiths. 2008. *Risk Assessment in Geotechnical Engineering*: New York: John Wiley & Sons, Inc.: 461 p.

Fiorotto, R. 2000. *Improvement of the Mechanical Characteristics of Soils by Jet Grouting*, personal communication.

Flora, A., G. P. Lignola, and G. Manfredi. 2007. A semiprobabilistic approach to the design of jet-grouted umbrellas in tunelling. *Ground Improvement* 11(4): pp. 207–217.

Flora, A. and S. Lirer. 2011. Interventi di consolidamento dei terreni, tecnologie e scelte di progetto. *Proceedings of the 24th National Conference of Geotechnical Engineering 'Innovazione tecnologica nell'Ingegneria Geotecnica'*, Napoli, Italy, June 22–24, 2011: pp. 87–148 [in Italian].

Flora, A., S. Lirer, and M. Monda. 2012a. Probabilistic design of massive jet-grouted water sealing barriers, In L. F. Johnsen, D. A. Bruce, and M. J. Byle, eds., *ASCE Proceedings of the 4th International Conference on Grouting and Deep Mixing* 2: pp. 2034–2043.

Flora, A., S. Lirer, G. P. Lignola, and G. Modoni. 2012b. Mechanical analysis of jet-grouted supported structures, In G. Viggiani, ed., *Proc. of the 7th Int. Symposium on Geotechnical Aspects of Underground Construction in Soft Ground*, TC28 IS Rome: London: Taylor & Francis Group, May 16–18, 2011: pp. 819–828.

Flora, A., G. Modoni, S. Lirer, and P. Croce. 2013. The diameter of single-, double-, and triple-fluid jet grouting columns: Prediction method and field trial results. *Géotechnique* 63(11): pp. 934–945.

Fruglietti, A., R. Roberi, and R. Tornaghi. 1989. Fondazioni a pozzo di un ponte realizzate mediante consolidamento del terreno. *Proceedings of the 17th National Conference of Geotechnical Engineering*, Taormina, Italy, April 26–28, 1989: pp. 217–229 [in Italian].

Garassino, A. L. 1983. Uso delle colonne in terra stabilizzata con jet grouting come elemento provvisionale. *Proceedings of the 15th National Geotechnical Conference 2*, Spoleto, Italy, May 4–6, 1983, Associazione Geotecnica Italiana: pp. 155–161 [in Italian].

Garassino, A. L. 1997. *Design Procedures for Jet Grouting*. Seminar organised by THL Foundation Equipment Pte., Ltd., and Gaggiotti Far East Singapore.

Garassino, A. L. and S. Palmoso. 1983. Prove di spinta orizzontale su colonne in terra stabilizzata. *Proceedings of the 15th National Geotechnical Conference 3*, Spoleto, Italy: pp. 154–161 [in Italian].

GI-ASCE. 2009. *Jet Grouting Guideline*. Geo Institute of ASCE, Grouting Committee–Jet Grouting Task Force, American Society of Civil Engineers: 29 p.

Guatteri, G., A. C. Doria, J. L. Kauschinger, and E. B. Perry. 1988. Advances in the construction and design of jet grouting methods in South America. *Proceedings of the 2nd International Conference on Case Histories in Geotechnical Engineering*, St. Louis, MO, June 1–5, 1988, Paper 5.32.

Guatteri, G., A. Koshima, and M. R. Pieroni. 2012. Challenges in the execution of jet grouting curtains at the Estreito HPP. *ASCE Proceedings of the 4th International Conference on Grouting and Deep Mixing*: pp. 2102–2111.

Haldar, A. and S. Mahadevan. 2000. *Probability, Reliability, and Statistical Methods in Engineering Design*: New York: John Wiley & Sons: 304 p.

Harlow, F. H. and P. I. Nakayama. 1968. *Transport of Turbulence Decay Rate*. Report of the Los Alamos Science Laboratory, LA-3854.

Heng, J. 2008. *Physical Modelling of Jet Grouting Process*. PhD thesis, University of Cambridge, Cambridge, United Kingdom: 260 p.

Hinze, J. O. 1975. *Turbulence*, 2nd ed.: New York, McGraw-Hill: 320 p.

Hong, W. P., D. W. Kim, M. K. Lee, and G. G. Yea. 2002. Case study on ground improvement by high-pressure jet grouting. *Proceedings of the 12th International Conference on Offshore and Polar Engineering*, International Society of Offshore and Polar Engineers, Kitakyushu, Japan, May 26–31, 2002: pp. 610–615.

Ichihashi, Y., M. Shibazaki, H. Kubo, M. Iji, and A. Mori. 1992. Jet grouting in airport construction, In *Grouting, Soil Improvement, and Geosynthetics*, ASCE Geotechnical Special Publication 30: pp. 182–193.

Iwasaki, M. 1989. Application of water jet for drilling rock undersea. *Turbomachinery* 17(12): pp. 761–767 [in Japanese].

JJGA. 2005. *Jet Grouting Technology: JSG Method, Column Jet Grouting Method*. Technical Information of the Japanese Jet Grouting Association, 13th ed. (English translation), October 2005: 80 p.

Katzenbach, R., A. Weidle, and H. Hoffmann. 2001. Jet grouting: Chance of risk assessment based on probabilistic methods, In M. T. Durgunoglu, ed., *Proceedings of the 15th ICSMFE*, Istanbul, Turkey, August 27–31, 2001: pp. 1763–1766.

Kauschinger, J. L., R. Hankour, and E. B. Perry. 1992. Methods to estimate composition of jet-grout bodies, In *Grouting, Soil Improvement, and Geosynthetics*, ASCE Geotechnical Special Publication 30: pp. 169–181.

Kutzner, C. 1996. *Grouting of Rock and Soil*: Rotterdam, Netherlands: Balkema: 271 p.

Laguzzi, G. and S. Pedemonte. 1991. Esperienze di jet-iniezione per l'esecuzione ed il consolidamento di linee ferroviarie nel compartimento di Milano. *Proceedings of the Conference 'Il Consolidamento del Suolo e delle Rocce nelle Realizzazioni in Sotterraneo'*, Milano, Italy, March 18–20, 1991, Società Italiana Gallerie, vol. 2, session C: pp. 209–217 [in Italian].

Lancaster, L. C. 2005. *Concrete Vaulted Construction in Imperial Rome: Innovations in Context*: New York: Cambridge University Press: pp. xxii + 274.

Langbehn, W. K. 1986. *The Jet Grouting Method: Application in Slope Stabilisation and Landslide Repair*: Master of Engineering Report, Department of Civil Engineering, University of Berkeley (California): 156 p.

Langhorst, O. S. 2012. Jet grouting foundation under the overpass of the A27 in the polder construction sealed with a foil at Amelisweerd. *Proceedings of the ISSMGE–TC211 International Symposium on Ground Improvement*, Brussels, Belgium, May 31–June 1, 2012: pp. 281–289.

Langhorst, O. S., B. J. Schat, J. C. W. M. de Wit, P. J. Bogaards, R. D. Essler, J. Maertens, B. K. J. Obladen, C. F. Bosma, J. J. Seluwaegen, and H. Dekker. 2007. Design and validation of jet grouting for the Amsterdam Central Station. *Geotechniek*, Special Number on Madrid 14th ECSMGE: pp. 20–23.

Leach, S. J. and G. L. Walker. 1966. Some aspects of rock cutting by high-speed water jets. *Philosophical Transactions of the Royal Society of London Series A*, 260(1100): pp. 295–308.

Lesnik, M. 2001. Methods to determine the dimension of jet-grouted bodies. *Proceedings of the 14th Young Geotechnical Engineers Conference*, Plovdiv, Bulgaria, September 15–19: pp. 363–371.

Lignola, G. P., A. Flora, and G. Manfredi. 2008. A simple method for the design of jet-grouted umbrellas in tunelling. *ASCE Journal of Geotechnical and Geoenvironmental Engineering* 134(12): pp. 1778–1790.

Lignola, G. P., A. Flora, and G. Manfredi. 2009. Effects of defects on structural safety of jet-grouted umbrellas in tunelling. *Proceedings of the 10th International Conference on Structural Safety and Reliability*, Osaka, Japan, September 13–17, 2009: Paper CH454.

Lunardi, P. 1992. Il consolidamento del terreno mediante jet grouting. *Quarry and Construction*, March 1992: pp. 127–140 [in Italian].

Maertens, J. and W. Maekelberg. 2001. Special applications of the jet grouting technique for underpinning works. *Proceedings of the 15th ICSMFE*, Istanbul, Turkey, August 27–31, 2001: pp. 1795–1798.

Mandolini, A. and V. Manassero. 2011. Interventi geotecnici di carattere strutturale. *Proceedings of the 24th National Conference of Geotechnical Engineering 'Innovazione Tecnologica nell'Ingegneria Geotecnica'*, Napoli, Italy, June 22–24: pp. 5–49 [in Italian].

Mauro, M. and F. Santillan. 2008. Large-scale jet grouting and deep mixing test program at Tuttle Creek Dam. *Proceedings of the 33rd Annual and 11th International Conference on Deep Foundations*, Paper 1607: 10 p.

Meinhard, K., D. Adam, and R. Lackner. 2010. Temperature measurements to determine the diameter of jet-grouted columns. *Proceedings of the International Conference on Geotechnical Challenges in Urban Regeneration*, London, United Kingdom, May 26–28: 8 p.

Miki, G. 1973. Chemical stabilization of sandy soils by grouting in Japan. *Proceedings of the 8th ICSMFE*, Moscow, Russia, June 1973: pp. 395–405.

Miki, G. 1982. The newest techniques on chemical grouting and jet grouting. *Proceedings of the Symposium on Recent Development in Ground Improvement Techniques*, Bangkok, Thailand, November 29–December 3, 1982: pp. 279–288.

Miki, G. and W. Nakanishi. 1984. Technical progress of the jet grouting method and its newest type. *Proceedings of the International Conference on In Situ Soil and Rock Reinforcement*, Paris, France, October 9–11, 1984: pp. 195–200.

Mitchell, J. K. and R. K. Katti. 1981. Soil improvement, state-of-the-art report. *Proc. of the X Int. Conf. Soil Mechanics and Found. Eng., Stockholm*, Vol. 4, Balkema, Rotterdam, June 1981: pp. 509–565.

Miyasaka, G., Y. Sasaki, T. Nagata, M. Shibazaki, M. Iji, and M. Yoda. 1992. Jet grouting for a self-standing wall, In *Grouting, Soil Improvement, and Geosynthetics*. ASCE Geotechnical Special Publication 30: pp. 144–155.

Modoni, G. and J. Bzówka. 2012. Design of jet grouting for foundation. *ASCE Journal of Geotechnical and Geoenvironmental Engineering* 138(12): pp. 1442–1454.

Modoni, G., J. Bzówka, A. Juzwa, A. Mandolini, and F. Valentino. 2012. Load-settlement responses of columnar foundation reinforcements. *Proceedings of the ISSMGE–TC211 International Symposium on Ground Improvement 3*, Brussels, Belgium, May 31–June 1, 2012: pp. 491–503.

Modoni, G., P. Croce, and J. Bzòwka. 2008b. Design of jet-grouted foundations for maritime structures. *Proceedings of the 11th Baltic Sea Geotechnical Conference: Geotechnics in Maritime Engineering*, Gdanks, Poland, September 15–18, 2008: pp. 989–996.

Modoni, G., P. Croce, and L. Mongiovì. 2006. Theoretical modelling of jet grouting. *Géotechnique* 56(5): pp. 335–347.

Modoni, G., P. Croce, and L. Mongiovì. 2008a. Theoretical modelling of jet grouting: Closure. *Géotechnique* 58(6): pp. 533–535.

Mohan, D., G. S. Jain, and D. Sharma. 1967. Bearing capacity of multireamed bored piles. *Proceedings of the 3rd Asian Conference on SMFE*, Haifa, Israel, May 1967: pp. 103–110.

Momber, A. W. and R. Kovacevic. 1997. Test parameter analysis in abrasive jet cutting of rock-like materials. *International Journal of Rock Mechanics and Mining Sciences* 34(1): pp. 17–25.

Mongiovì, L., P. Croce, and A. Zaninetti. 1991. Analisi sperimentale di un intervento di consolidamento mediante gettiniezione. *Proceedings of the 2nd National Conference of the Researchers of Geotechnical Engineering*: pp. 101–118 [in Italian].

Morey, J. and D. W. Campo. 1999. Quality control of jet grouting on the Cairo metro. *Ground Improvement* 3(2): pp. 67–75.

Nakanishi, W. 1974. *Method for Forming an Underground Wall Comprising a Plurality of Columns in the Earth and Soil Formation*, U.S. Patent 3,800,544: 8 p.

Nanni, E., J. Oberhuber, and P. Froldi. 2004. L'utilizzo della metodologia jet grouting per l'ampliamento e la ristrutturazione di edifici esistenti (in ambito urbano). *Proceedings of the 22nd National Geotechnical Conference*, Palermo, Italy, September 2004, Associazione Geotecnica Italiana: pp. 403–407 [in Italian].

Neville, A. M. and J. J. Brooks. 1987. *Concrete Technology*: Harlow, United Kingdom: Pearson Education, Ltd.: 431 p.

Nikbakhtan, B. and M. Osanloo. 2009. Effect of grout pressure and grout flow on soil physical and mechanical properties in jet grouting operations. *International Journal of Rock Mechanics and Mining Sciences* 46: pp. 498–505.

Panet, M. and A. Guenot. 1982. Analysis of convergence behind the face of a tunnel. *Tunnelling*: The Institution of Mining and Metallurgy: pp. 197–204.

Parry Davies, R., R. M. H. Bruin, G. Wardle, and M. G. Nixon. 1992. Stabilization of pier foundation using jet grouting techniques, In *Grouting, Soil, Improvement and Geosynthetics*, ASCE Geotechnical Special Publication 30: pp. 156–168.

Peck, R. B., W. E. Hansen, and T. H. Thornburn. 1953. *Foundation Engineering*: New York: John Wiley & Sons: 401 p.

Perelli Cippo, A. and R. Tornaghi. 1985. Soil improvement by jet grouting, In S. Thorburn and J. F. Hutchison, eds., *Underpinning*: Surrey, United Kingdom: Surrey University Press: pp. 276–292.

Pinto, A., R. Tomàsio, X. Pita, and A. Pereira. 2012. Ground treatment solutions using jet grouting. *ASCE Proceedings of the 4th International Conference on Grouting and Deep Mixing*: pp. 2112–2121.

Popa, A. 2001. Underpinning of buildings by means of jet-grouted piles. *Proceedings of the 4th International Conference on Ground Improvement*, Helsinki, Finland, September 4–6: pp. 221–227.

Przyklenk, K. and M. Schlatter. 1986. Simulation of the cutting process in water jetting with the finite element method. *Proceedings of the 8th International Symposium on Jet Cutting Technology*, Paper 12, BHRA Fluid Engineering, Durham, England: pp. 125–136.

Raffle, J. F. and D. A. Greenwood. 1961. The relationship between the rheological characteristics of grouts and their capacity to permeate soils. *Proc. of the 5th Int. Conf. Soil Mech. & Found. Engng.* 2: pp. 789–793.

Russo, G. and G. Modoni. 2005. Monitoring results of a tunnel excavation in urban area. *Proceedings of the 5th International Symposium of the Technical Committee TC28: Geotechnical Aspects of Underground Construction in Soft Ground*, Amsterdam, Netherlands, June 15–17, 2005: pp. 751–756.

Ryjevski, M., K. Shramko, and E. Baghirova. 2009. Jet grouting application for Dubai Metro Construction. *Proceedings of the World Tunnel Congress*, Budapest, Hungary, May 23–28, 2009, Paper O-10-19: 10 p.

Saglamer, A., R. Duzceer, A. Gokalp, and E. Yilmaz. 2001. Recent applications of jet grouting for soil improvement in Turkey. *Proceedings of the 15th ICSMGE*, Istanbul, Turkey, August 27–31: pp. 1839–1842.

Sanella, A. 2007. *Soil Improvement Process Using Jet Grouting*. European Patent EP 1-795-655-A1: 5 p.

Santoro, V. M. and B. Bianco. 1995. Realizzazione di una struttura in gettiniezione per il sostegno di uno scavo in depositi piroclastici: Controllo di qualità in corso d'opera. *Proceedings of the 19th National Geotechnical Conference*, Pavia, Italy, September 19–21, 1995, Associazione Geotecnica Italiana: pp. 467–477.

Saurer, E., T. Marcher, and M. Lesnik. 2011. Grid space optimization of jet grouting columns, In A. Anagnostopoulos et al., eds., *Proceedings of the 15th European Conference on Soil Mechanics and Geotechnical Engineering*: Amsterdam, Netherlands: IOS Press: pp. 1055–1060.

Schmertmann, J. H. 1970. Static cone to compute static settlements over sand. *ASCE Journal of Soil Mechanics and Foundation Division* 96(3): pp. 1011–1043.

Sembenelli, P. and G. Sembenelli. 1999. Deep jet-grouted cutoffs in riverine alluvia for Ertan cofferdams. *Journal of Geotechnical and Geoenvironmental Engineering* 125(2): pp. 142–153.

Shen, S. L., C. Y. Luo, X. C. Xiao, and J. L. Wang. 2009. Improvement efficacy of RJP method in Shanghai soft deposit: Advances in ground improvement—Research to practice in the United States and China, In J. Han, G. Zheng, V. R. Schaefer, and M. S. Huang, eds., *Geotechnical Special Publication* 188: pp. 170–178.

Shibazaki, M. 1996. State-of-the-art grouting in Japan: *Grouting and Deep Mixing* 2: 851–867.

Shibazaki, M. 2003. State of practice of jet grouting, In L. F. Johansen, D. A. Bruce, and M. J. Byle, eds., *Proceedings of the 3rd International Conference on Grouting and Ground Treatment*, Vol. 1, ASCE Geotechnical Special Publication 120: pp. 198–217.

Shibazaki, M. and S. Ohta. 1982. A unique underpinning of soil solidification utilizing super high-pressure liquid jet. *Proceedings of the Conference of Grouting in Geotechnical Engineering*, pp. 680–693.

Shibazaki, M., S. Ohta, and H. Kubo. 1983. *Jet Grouting*: Tokyo, Japan: Kajima Publishing Company: pp. 15–20.

Shibazaki, M., M. Yokoo, and H. Yoshida. 2003. Development of oversized jet grouting, In L. F. Johansen, D. A. Bruce, and M. J. Byle, eds., *Proceedings of the 3rd International Conference on Grouting and Ground Treatment*, Vol. 1, ASCE Geotechnical Special Publication 120: pp. 294–302.

Shibazaki, M., H. Yoshida, and Y. Matsumoto. 1996. Development of a soil improvement method utilizing cross jet, In R. Yonekura and M. Shibazaki, ed., *Grouting and Deep Mixing*: Rotterdam, Netherlands: Balkema: pp. 707–710.

Sondermann, W. and P. S. Tóth. 2001. State of the art of the jet grouting shown on different applications. *Proceedings of the 4th International Conference on Ground Improvement*, Helsinki, Finland, June 7–9, 2001, Finnish Geotechnical Society: pp. 181–194.

Sopeña Mañas, L. M., C. Heras Meco, and J. Luis Rojo. 2001. Stabilisation works in a singular slope by jet grouting ground improvement. *Proceedings of the 4th International Conference on Ground Improvement*, Helsinki, Finland, June 7–9, 2001, Finnish Geotechnical Society: pp. 195–202.

Soyama, H., Y. Yanauchi, K. Sato, T. Ikohagi, R. Oba, and R. Oshima. 1996. High-speed observation of ultrahigh speed submerged water jets: *Experimental Thermal and Fluid Science* 12: pp. 411–416.

Stein, J. and J. Graße. 2003. Jet grouting tests and simulation, In Vanicek et al., eds., *Proceedings of the 13th ECMSGE*, Prague, Czech Republic, August 25–28, 2003: pp. 899–902.

Summers, D. A. 1995. *Waterjetting Technology*: London, United Kingdom: E & FN Spon: 882 p.

Tarricone, P. 1994. Jet grouting gains: *Civil Engineering (ASCE)* 64(12): pp. 40–43.

Terzaghi, K. and R. B. Peck. 1948. *Soil Mechanics in Engineering Practice*: New York: J. Wiley & Sons: 549 p.

Timoshenko, S. P. and J. N. Goodier. 1987. *Theory of Elasticity, 3rd ed.*: New York: McGraw-Hill: 416 p.

Tornaghi, R. 1989. Trattamento colonnare dei terreni mediante gettiniezione (jet grouting). *Proceedings of the 17th National Conference of Geotechnical Engineering*, Taormina, Italy, April 26–28: pp. 193–203 [in Italian].

Tornaghi, R. 1993. Controlli e bilanci analitici dei trattamenti colonnari mediante jet grouting. *Rivista Italiana di Geotecnica* 93(3): pp. 217–234 [in Italian].

Tornaghi, R. and A. Perelli Cippo. 1985. Soil improvement by jet grouting for the solution of tunelling problems. *Proceedings of the Conference on Tunnelling*, Brighton, United Kingdom: pp. 265–275.

Tornaghi, R. and A. Pettinaroli. 2004. Design and control criteria of jet grouting treatments. *Proceedings of the International Symposium on Ground Improvement*, ASEP-GI 2004: Paris, France: Ecole Nationale des Ponts et Chaussées: pp. 295–319.

USBR. 1981. *Concrete Manual: A Manual for the Control of Concrete Constructions, 8th ed.*: Denver, CO: U.S. Bureau of Reclamation: 627 p.

van der Stoel, A. E. C. 2001. *Grouting for Pile Foundation Improvement*: PhD thesis, Delft University Press: 217 p.

van Tol, F. 2004. Lessons learned from jet grouting at a tunnel project in the Hague. *Proceedings of the International Symposium on Ground Improvement*, ASEP-GI 2004 – Ecole Nationale des Ponts et Chaussées – Paris, France: pp. 321–331.

van Tol, A. F., A. J. E. van Riel, and J. K. Vrijling. 2001. Towards a reliable design for jet grout layers. *Proceedings of the ICSMGE* 15(3): pp. 1887–1890.

Vesic, A. S. 1963. *Bearing Capacity of Deep Foundation in Sand*. Highway Research Board Record No. 39: 153 p.

Vesic, A. S. 1975. Bearing capacity of shallow foundations, In M. F. Winterkorn and M. Y. Fang, eds., *Foundation Engineering Handbook*: Van Nostrand Reinhold.

Viggiani, C., A. Mandolini, and G. Russo. 2011. *Piles and Pile Foundations*: CRC Press, Taylor & Francis: 278 p.

Wang, J. G., B. Oh, S. W. Lim, and G. S. Kumar. 1999. Effect of different jet grouting installations on neighbouring structures, In C. F. Leung, S. A. Tan, and K. K. Phoon, eds., *Field Measurements in Geomechanics*: Rotterdam, Netherlands: Balkema: pp. 511–516.

Wang, Z. F., S. L. Shen, and J. Yang. 2012. Estimation of the diameter of jet-grouted column based on turbulent kinematic flow theory. *Proceedings of the Conference on Grouting and Deep Mixing* 2, ASCE Geotechnical Special Publication 228: pp. 2044–2051.

Wanik, L. and G. Modoni. 2012. Numerical analysis of the diffusion of submerged jets for jet grouting application. *Incontro Annuale dei Ricercatori di Geotecnica (IARG)*, Paper 3.B.8: 6 p.

Wright, S. J. and L. C. Reese. 1977. *Construction of Drilled Shaft and Design for Axial Loading; Manual on Drilled Shaft 1*: Washington, DC: U.S. Department of Transportation: 140 p.

Xanthakos, P., L. W. Abramson, and D. A. Bruce. 1994. *Ground Control and Improvement*: New York: John Wiley & Sons, Inc.: 670 p.

Yahiro, T. and H. Yoshida. 1973. Induction grouting method utilizing high-speed water jet. *Proceedings of the 8th ICSMFE*, Moscow, Russia, June 1973: pp. 402–404.

Yahiro, T., H. Yoshida, and K. Nishi. 1974. Soil improvement utilizing a high-speed waterjet and air jet. *Proceedings of the 6th International Symposium on Water Jet Technology*, Cambridge, United Kingdom, Paper J62, August 30–31: pp. 397–428.

Yahiro, T., H. Yoshida, and K. Nishi. 1975. The development and application of Japanese grouting system. *Water Power Dam Construction* 27: pp. 55–69.

Yahiro, T., H. Yoshida, and K. Nishi. 1982. Soil improvement method utilizing a high-speed water and air jet on the development and application of columnar solidified construction method. *Proceedings of the 6th International Symposium on Jet Cutting Technology*, United Kingdom: University of Surrey: pp. 397–427.

Yahiro, T., H. Yoshida, and K. Nishi. 1983. *Soil Modification Utilizing Water Jet*: Tokyo, Japan: Kajima Publishing Company: pp. 7–20.

Yoshida, H. 2012. Recent development in jet grouting, In L. F. Johansen, D. A. Bruce, and M. J. Byle, eds., *Proceedings of the 4th International Conference on Grouting and Deep Mixing* 2, ASCE Geotechnical Special Publication 228: pp. 1548–1561.

Yoshida, H., S. Jimbo, and S. Uesawa. 1996. Development and practical applications of large diameter soil improvement method, In R. Yonekura and M. Shibazaki, eds., *Proceedings of the Conference on Grouting and Deep Mixing*, Tokyo, Japan: Balkema, May 14–17, 1996: pp. 721–726.

Yoshitake, I., T. Mitsui, T. Yoshikawa, A. Ikeda, and K. Nakagawa. 2003. An evaluation method of ground improvement by jet grouting. *Proceedings of the Japan Society of Civil Engineers* 735: pp. 215–220.

Young, W. C. and R. G. Budynas. 2002. *Roark's Formulas for Stress and Strain*: New York: McGraw Hill: 676 p.

Yu, F. C. 1994. *Mechanical Properties of Jet-Grouted and Deep Mixed Soilcrete*: Master's thesis, Department of Civil Engineering, National Chiao Tung University, Hsinchu, Taiwan: 160 p [in Chinese].

Index

Printed in the United States
by Baker & Taylor Publisher Services